To Professor R T Rainoke

Chief Exec

E P S R C

from

Hiroaki Yanagida

1995-5-31

TECHNOLOGY'S NEW HORIZONS

TECHNOLOGY'S NEW HORIZONS

Conversations with Japanese Scientists

Edited by

HIROAKI YANAGIDA

Department of Applied Chemistry
University of Tokyo

Oxford   New York   Tokyo
OXFORD UNIVERSITY PRESS
1995

Oxford University Press, Walton Street, Oxford OX2 6DP

Oxford   New York

Athens   Auckland   Bangkok   Bombay
Calcutta   Cape Town   Dar es Salaam   Delhi
Florence   Hong Kong   Istanbul   Karachi
Kuala Lumpur   Madras   Madrid   Melbourne
Mexico City   Nairobi   Paris   Singapore
Taipei   Tokyo   Toronto
and associated companies in
Berlin   Ibadan

Oxford is a trade mark of Oxford University Press

Published in the United States
by Oxford University Press Inc., New York

Parts of the interviews appearing in this book first appeared in Japanese in a series of books
published by Mita Press, Tokyo.

Oxford University Press is grateful to the Daido Life Foundation for financial support of the
translation and editing of this book.

A catalogue record for this book is available from the British Library

Library of Congress Cataloging in Publication Data
Yanagida, Hiroaki, 1935
Technology's new horizons: conversations with Japanese scientists
/Hiroaki Yanagida 1. Technological innovations.   2. Technology–Philosophy.
3. Sciences–Philosophy   4. Scientists–Japan   I Title.
T173.8.Y35   1995   601–dc20   94–47078
ISBN 0 19 856514 3

Typeset by EXPO Holdings, Malaysia

Printed in Great Britain by Biddles Ltd, Guildford & King's Lynn.

FOREWORD

PROFESSOR RICHARD L. GREGORY, UNIVERSITY OF BRISTOL

These interviews with Japanese engineers are a revelation, at least for Western readers. Each interview gives a résumé of the distinguished individual's background interests and working philosophy. There are, also, many suggestive comments on the Japanese view of their culture, and its relations with Europe and America. One senses the growing confidence since the last World War of the Japanese ability to initiate, as well as improve and produce products with technical excellence. It is hard to resist a Japanese car! It just keeps on going and going.

Takanori Okoshi from reading philosophers such as Kant, Schopenhauer, and Nietzsche, turned to seeing the light for optical communication technology. Realizing that existing microwave technology was of the wrong scale, he transferred its technology to fit the wavelengths of light. This was associated with the pioneering work of the Bell Telephone Laboratories in America, but there is no doubt that the Japanese contribution was fundamentally important, and is having incredible impact on human and computer world communication.

Yasuharu Suematsu contributed to the essential source of optotechnology systems—suitable lasers. There is a fascinating account here of matching the laser to fibres and/or matching fibres to available lasers. One senses that these decisions (sometimes made from anticipated but not yet proved discoveries) have immense consequences financially, and for the long term future. It must be a tremendous headache to work with rapid innovation and at the same time produce effective state-of-the-art devices. If only the public appreciated these battles, wrestling with Nature—which are surely more exciting than war games or real wars. Yet, Yasuharu Suematsu says that: 'It may surprise many Westerners that, even in Japan, technology is by no means regarded as the basis of civilization.... Engineers are simply seen as people who beaver away making things. The leading figures in our society are lawyers and economists who see technology as only another way of making money.' This is worth pondering.

Hioroyuki Sakaki leads us to 'the last frontier of electronics'. This is incredibly dramatic, as it leads to an ultimate goal of physics. He describes the restricting of electrons from three dimensions to two, to one—then envisages how a single electron could be captured and imprisoned in a minute box—so far too small even for Japanese fingers to make! This possibility has come from his discovery of quantum effects at room temperature. Here we see fundamental physics and state-of-the-art technology converging, to meet at a point that no one can predict. It may allow computers to read with the smallest possible unit, single electrons; it may reveal new quantal principles when the electron is contained in its box. It is very clear that Hioroyuki Sakaki, and indeed I would think all the subjects of this book, are primarily motivated not by money, but rather by dreams of philosophical and physical understanding, together with how new insights may be applied, hopefully, for human benefit.

Makoto Nagao has developed the most impressive machine translation at present available, between English and Japanese. He started from frankly mystical interests: the seventh-century mythical history of Japan, the *Kojiki*, and the *Manyoshu*, ancient Japanese poetry compiled in the eighth century. The Editor comments: 'It strikes me that language can be approached from the point of view of computers, and at the same time as a cultural, semantic, or philosophical problem.' Again and again the overlap between science and the humanities emerges in this revealing book.

Kazuhiro Fuchi was President of the Institute for New Generation Computer Technology (ICOT). This project for developing the 'Fifth Generation Computer' was terminated in 1993. It might be seen as analogous to the American commitment to land on the moon within ten years—except that the moon landing was achieved within target (1969) and the Fifth Generation Computer has not yet arrived. Is this simply a failure? Kazuhiro Fuchi responds to two levels of criticism. He rejects the media's account that the major problems of artificial intelligence (AI) research would be solved within ten years, saying that no-one at ICOT made such a claim. And it was not expected that, for example, a machine translation system would have the capabilities of humans. Secondly, ICOT was set up to experiment with new ideas, rather than produce practical devices. Clearly there is considerable ambiguity in the history of ICOT—as well, indeed, as in the history of AI in America and Britain. (Perhaps this is typical of human history: it is written after the victory and by the victors; but these hardly exist so far in AI.) The essential plan was to develop parallel inference. This required new hardware and new software. It was influenced by concepts of neural computing of the brain, including neural nets. The performance of AI is compared with the brain, which has been developing for many millions of years. If the brain was understood better there could be more effective transfer of technology from protoplasm to silicon. This is a multidisciplinary, completely international enterprise in which we may see ourselves as marrying the products of our intelligence, so the future may be indissolubly linked brain–machine intelligence.

Shun-ichi Amari bases his ideas on AI very much on the brain seen as a neural computer. He spent a lot of time considering whether the brain works in the same way as a computer. As he says, 'It's interesting that while the computer can be described perfectly as a Turing machine, the brain, although we live with it and have an intuitive grasp of its workings, has so far revealed nothing about the principles of its own operation.' It could be added that it is current technologies, especially of brain scanning and computer analysis of complex data, that are now revealing brain processes at a rapidly increasing rate. Shun-ichi Amari's approach was to start with random circuits, then look at neurons arranged spatially in layers, and then search for the kinetics of the pattern by which 'the excitement of one neuron is transmitted to others around it. Then, by way of basic research on the learning process, we investigated possible models for memory where the links between neurons change as a result of this action.' This is very much in the spirit of the pioneering Canadian psychologist, Donald Hebb; an approach which went out of fashion with the ascendancy of digital computing, but has now returned. Shun-ichi Amari emphasizes what he calls the 'form' approach: that the brain is not only doing simple processing (like a computer) but perhaps more important, it processes objects in perception 'intuitively and unconsciously as form. Also many items related to but not actually represented in symbols— like the raw material and even the history [of a seen object]—are activated at the same time as the brain grasps [the object] as a form.' Asked whether he sees himself as a leader in the search to unite brain and computer, he replies: 'No, maybe just a cheer-leader—and, I would add, don't follow your leader blindly!'

Hiroaki Yanagida wants to design intelligence into materials. He is especially interested in ceramics. His point is that intelligence can exist outside brains: the immune system copes with all manner of problems adaptively, without bothering the brain and without consciousness. He sees materials capable of self-diagnosis—so warning of corrosion or external threat. He sees humidity sensors, comparable to human skin, and self-adjusting materials such as photochromic glass, but more sophisticated. As an example: 'a "wise" material for the next generation might be traditional Japanese wood. Wood as a structural material adjusts to Japan's very humid environment by its capacity for self-diagnosis and self-recovery mechanisms without any help from micro-chips! I think it would be marvellous to find a new material which not only retains the merit of wood but also overcomes its weaknesses. Such a material, of which I dream, should be considered neither smart nor intelligent but "wise", in the sense that it's full of wisdom.' And he proposes that materials should be breakable for recycling, and as simple as possible to promote techno-democracy.

Hiroyuki Yoshikawa is President of the University of Tokyo, so his interests range from engineering to issues of education. He pioneered the concept of 'general theory of design'. He is one of the few to ask 'What is engineering?' although many have asked 'What is science?' His answer includes the notion that while science deals with infinite possibilities, engineering is always limited to finite solutions from a set of restricted possibilities. A key issue is how human beings classify the world. For example there are three sorts of meat: one fresh, one rotten, the last dry. A primitive man or woman will examine each option, and after experimenting will classify them. 'From a viewpoint of edibility, the fresh meat stands out from the others. However, the human soon notices that the fresh meat will also rot eventually. Thus, from the viewpoint of durability, fresh meat and rotten meat are similar, while dried meat stands apart. It can't be eaten but it won't change in time. So a primitive person begins to wonder if there is a kind of meat which is both edible and durable even though he doesn't have it in front of him.' Isn't pemmican edible dried meat? However this may be, the notion is that creative design follows from existing designs and is limited to generally rather few possibilities. This also applies to biological 'design' in the evolution of species. Professor Yoshikawa is attracted to genetics, as the engineer, also, has to design from what already exists. Again related to biology: 'No one expects a man of seventy to behave as though he was thirty. But we are demanding that the nuclear reactor doesn't age or die.'

Isao Karube develops biological sensors for electronic devices. These started using enzymes (protein molecules) and then using micro-organisms. The BOD (biological oxygen demand) sensor is used for environmental monitoring of polluted rivers. This electronic–biology link up leads to possibilities of introducing electronics circuits into the damaged nervous system. 'My wildest dream is to create a form of biocommunication using sensors that measure energy at a different dimension, such as quantum wave sensors. Of course, if it becomes possible to put sensors easily into the brain, I suppose psychological activities such as memory, thought, and emotion will be sensed electronically.' And: 'If it becomes possible to sense the flow of transmitters in what is regarded as a modern disease such as a manic-depressive psychosis or stress, we could view these diseases using objective data. The day might come when we could decide whether or not to work by observing the dynamic flow of transmitters in the brain! By integrating sensors for different transmitters, a "communication cap" could be made which would enable us to communicate with others without using language!' Isao Karube comments generally: 'The tendency to think of Japanese technology as being without originality is completely outdated. Today, the technology of Japan, the US, and Europe is almost at the same level, and the problems they are facing have many things in common. To put it very simply, I think the essence of these common problems is how to transform the materialistic civilization that is characteristic of the twentieth century to a spiritual civilization in the twenty-first century. Recent research, for example, into the electronic interface with the brain or chemical communication with plants, is a way that we technologists, worldwide, have of changing our materialistic civilization.'

Teruhiko Beppu working in biotechnology, searches among microorganisms for functions which might work to our advantage. Such secrets of Nature may be applied quite directly, and they may suggest artificially created molecules, or even new organisms. Teruhiko Beppu thinks that it was beyond human technology to have invented penicillin: essentially by chance Fleming discovered something of immediate use which led to 'drug designs' which might never occur in Nature. So here again technology is intimately associated with biology, and can go beyond the solutions of natural selection. For him, research into penicillin is an example of 'the need to go beyond basic research, testing out alternative applications. I think this is a very good example of a discovery in applied research leading to a new theory in pure science.' He goes on: 'The history of science is full of such cases. A famous example occurred in the Sony Laboratories, where the tunnel effect was discovered by chance when they were trying to develop new transistors.'

Science is the most international of all human activities; yet there are limits. Teruhiko Beppu says: 'I graduated in 1956, just when Watson and Crick's model [of DNA and its significance] was presented, but I knew nothing about it. In Japan, for a while, this model was not even taught in postgraduate lectures.' Do we, in the West, know enough of Japanese discoveries? This book may help to open our eyes.

These conversations with Japanese engineers are rich in the usually unstated wisdom of technology, which in fact is shaping our world. It is very strange that the methodology and aims of science are extensively discussed— especially by philosophers—but the same is not so for technology. Yet, as Teruhiko Beppu says: 'These are good examples of technology leading science. It is a myth that we always move from pure to applied science. A true investigation begins only when we confront an unexpected situation. It is only half true that science always precedes technology.'

It is not clear what prestige engineers have in Japan. I suspect it is higher than in the UK. But wouldn't it be a huge advance in our social, scientific, and technological wisdom, if engineers and technologists realized to the full their enormous contribution to the great adventure of discovering the nature of the universe and ourselves? Wouldn't this appreciation give Western industry the meaning and indeed glamour it sadly lacks? For pure science is utterly dependent on the devices and techniques of technology. Indeed they push back the frontiers of metaphysics, as new experiments become possible. Where would we be without giant telescopes and microchips? Pure mathematicians look to computers for proof—even preferring the computer's machinations to their own intuitive 'Aha!' feeling that they are right. In ancient Egypt, the weighing of the soul upon death for judgement leading to their heaven or hell, was not trusted to man or god. Thoth, the God of Wisdom, merely ensured that the scales of justice were kept calibrated. It was the machine that made the moral decision. It isn't only that machines decide moral issues (such as who should receive hospital treatment) but as technology advances so new moral dilemmas are created: for as it becomes possible to achieve more it must be decided what *should* be done from what *can* be done. Will these be future machine decisions?

It is impossible to read this book without sensing the importance of these issues, and realizing that it is the countries with industrial wealth who are now leading the world of ideas—as indeed was so in the seventeenth century, when England and Holland, together with the special imagination of Italy, introduced the enlightenment of post-Aristotelian physics.

As our bodies and brains were designed by blind trial-and-error of natural selection, the sole goal at each step being survival, so the technology determining our future proceeds by blind processes to unknown unforeseeable futures. Listening to these philosopher–engineers living at the edge, indeed creating the edge, of present possibilities, is perhaps the nearest we can come to seeing the future. Whether this understanding makes it possible to choose what to invent and control what lies ahead is another question.

CONTENTS

INTRODUCTION

HIROAKI YANAGIDA

For this book, leading technologists working in a variety of specialities were asked to talk about their work. They were also asked about their own lives, the background to their research, their views about society and education, religion and philosophy, and even about their personal hobbies. These discussions are as diverse as can be imagined and it seems impossible to draw any systematic conclusions from them. The book can be seen as a preliminary step to deeper and broader discussions in the future. An attempt was made to discuss the current status of their research for the non-specialist reader. As a result, this book contains the most advanced discussion about several technologies which can be viewed as being of the utmost importance to contemporary civilization such as optical communications, computers, new materials, and biotechnology. In addition, this book will be a good introduction to the fascinating fields of advanced technology for students on engineering courses.

It may be a surprise for Western readers that much of this research was initiated in Japan, because, according to the familiar stereotype, Japanese technology has always been practical rather than theoretical. Technological products from Japan have dominated the world because the Japanese people have a completely pragmatic caste of mind. This stereotype is not completely wrong; indeed it is almost true. In fact, after reading these interviews, Western readers may be surprised that so little of the discussion is theoretical. Japanese technologists are usually down-to-earth even when they are asked to be philosophical; they seem ashamed to speculate. For this reason we were obliged to offer them alcohol to loosen their tongues! It should be emphasized also that the scientists interviewed here are not stereotypically Japanese. They are pioneers with highly original minds, who have sometimes suffered from the resistance which is rooted in a Japanese culture hostile to change. Japan can be seen as the last industrialized country in which a call for a philosophy of technology is heard.

However, this is not the true picture. Apart from the fact that there has been a very strong tradition of philosophical thought in Japan, there are very strong cultural reasons why we Japanese people have had to pose questions about technology. It should be remembered that modern technology is the twin brother of modern science, which is rooted deeply in Western civilization. Modern technology is, it could be said, a purely Western phenomenon. In fact, it has matured in the West over several centuries beginning with giants like Leonard da Vinci. It is a natural outcome of the growth of Western civilization. It cannot be denied that there is a Western flavour in every aspect of contemporary technology, from computers and robots to rocket missiles and space shuttles. To the Japanese, technology seems like an unnatural transplant. Though there were strong craft traditions before the Meiji Restoration, technologies based on the science of Gallileo, Descartes, and Newton were alien to Japan's cultural identity. We have been forced to adapt quickly to something quite foreign to our traditions in order to survive in the cruel competitive world of the modern superpowers. After a hundred and thirty years of restless endeavour, we have now apparently assimilated Western technology. It is aptly stated by one interviewee that we are now 'more Western than the West'. Because it is somewhat unreal for us, however, technology in Japan sometimes goes too far. This situation is exemplified in the Japanese landscape with its labyrinthine network of electricity cables, so different from the country of Hokusai and Hiroshige, and this is the target of severe criticism by Professor Okoshi, the renowned leader in optical communications and the first interviewee in this book. Also Japan is one of those advanced countries in which environmental pollution is particularly high. Technologists have struggled with environmental problems in this country for years. I would maintain therefore that Japan is the country where we should enquire most seriously into the status of technology as a cultural phenomenon. I hope we can stimulate the Western reader with the unexpectedness of our ideas.

Apart from this general intention, this book has one specific purpose. We have often been told by Westerners that the Japanese are 'faceless'. This could be because the Western people exhibit their faces too much, while there has been little effort on the part of the Japanese to display their individual thoughts to the West. This has been especially true of the technologists who were responsible for the world-renowned technology of Japan. Therefore, the specific purpose of this book is to present for the first time the real individual voices of some of the leading Japanese technologists. In saying that, I do not mean that these technologists are speaking as the representatives of Japan. They have been selected because they are original and interesting people. Their thinking is by no means conventionally Japanese. Their work has universal merit. Although their subject matter is not specifically Japanese, many of their views, sometimes expressed unconsciously, do have a Japanese orientation. Many of the interviewees express critical views, for example about research in the USA. It cannot be denied that we are nationalistic to a degree but it should be pointed out that many interviewees are critical of the Japanese system itself. None of the technologists interviewed are chauvinists.

As this book is the first of its kind, I have tried to make it a valuable record of the history of technology. Care was taken, in this respect, in the interviews concerning optical communications. It is not generally acknowledged that twentieth-century civilization would not have been what it is without telecommunication technology. The replacement of electrical by optical communication is one of the most significant changes to have occurred in modern civilization. For example, the recent passion in the United States for the 'Information Superhighway' and for multimedia was only made possible by optical communications technology. Japan's highly significant contribution to this revolution might become the subject for research in the history of technology. One interview deals with a subject which has had much coverage in the media: the New Generation Computer Project of Japan. An institution called ICOT was created purely for this project. Although interesting in many other respects, the project gained historical significance because it was the first national technological project in Japan which could be described as truly adventurous. I have no difficulty with the stereotypical criticism from abroad that the Japanese are much better at refining the ideas of others than creating their own. It must be stated, however, that this is changing. ICOT is the first venture to meet this criticism. As any adventurous undertaking is likely to fail, it was reasonable that many foreign newspapers pictured it as a failure. The reader can decide whether this was really the case after reading the interview with Professor Fuchi, the former President of ICOT.

It might be helpful to give the reader some background information about the system which sustains Japanese technology. Roughly speaking, the healthy state of technology in Japan is due to the efforts of a vast number of engineers working in numerous industries. The technological strength of Japan is based upon an infinite number of inventions, improvements, and innovations which have taken place in industry, especially since the Second World War. As has often been pointed out, these achievements have been characteristic of Japanese perfectionism and collectivism, thus typifying our 'superb mediocrity'. Research in industry has been orientated very strongly towards application; it is only recently that the need for 'basic' research in industry has been recognized. There were no research institutes like Bell or IBM in Japan, though this situation is changing. Thus, basic technological research has been mostly carried out in our universities. British and American readers may not be fully aware that in Japan most of our national universities have the capacity for advanced research. These universities were originally created by the Meiji government to import ideas in order to catch up with Western civilization. Therefore, the basic tendency has been for Japanese national universities to refine and disseminate imported knowledge. This forms the basis of the notorious 'copy-cat' image of modern Japanese culture. Nevertheless, although this tendency was quite strong in our culture, the originality of Japanese research bloomed unexpectedly early in science and technology. It should be remembered that the first official department of engineering in a university was established in Tokyo. After only thirty years of the study of Western physics, Nagaoka's solar model of atomic structure was first conceived in Japan. The quiet invention of antennae by Yagi, ignored in Japan, revolutionized the war and led to the defeat of Yagi's

mother country. Since the Second World War, the imperial universities in
Japan have gradually changed, the most important aspect being the influence
of the USA. Readers will find evidence of this influence throughout these
interviews. Almost all of the technologists interviewed have had the experience of staying in America when they were young. The reader might be
interested to compare the interviews with Professor Okoshi and Professor
Sakaki, who were in the USA at almost the same time but at different ages.
Therefore, it can be said that the leaders in Japanese technology have been
deeply affected by American thinking and attitudes. It is a point of great
interest to see how Western thought interacts with the Japanese mind. On
the other hand, it is ironic that the spirit of higher education in Japan has not
changed very much since the war. Although academic freedom is encouraged, our national universities are constrained financially by the Ministry of
Education. Because its main aim has been the wide dissemination of standard
education, the Ministry of Education has not had a coherent policy concerning research in universities. Out of the total expenditure on research and
development, which is estimated at 3.5 per cent of the gross national product,
only one-fifth comes from the public sector and is mainly directed at the
universities, the rest being investment by industry aimed primarily at the
development of its own commercial products. Foreign visitors are astonished
to see poorly equipped laboratories in the most prestigious universities in
Japan. This situation is discussed by two Presidents of leading universities,
Professor Yoshikawa of the University of Tokyo and Professor Suematsu of
the Tokyo Institute of Technology. There have been efforts by the Ministry
of International Trade and Industries (MITI) to support research in advanced
technology, of which ICOT is an example. Recently transnational and international research groups have been formed using government support.
Therefore the financial restraints on the growth of original research seem to
be disappearing gradually. There are, however, still many problems in achieving originality in Japanese technological research. For example, we have been
criticized by foreign researchers for having little tradition of peer review,
without which the sensible allocation of the research budget cannot be
attained. This can be boiled down to the characteristic social position in
Japan: the emphasis on consensus and the repression of divergent opinions.
Many of the technologists interviewed have views on this subject.

   Finally, I would like to explain how this book came about. Originally, a
series of books called *Steering* were published in Japanese by Mita Press in
1989. This series was made up of interviews with thirty leading technologists
of Japan. Its main aim was to present to engineers, managers, and students the
real thoughts of cutting-edge technologists in Japan. Also it was intended to
explain the current state of advanced technologies. After this series had been
welcomed by the Japanese public, we were approached by Oxford University
Press about the possible publication of an English version. As the Japanese

version was not aimed at a foreign audience, extensive reorganization and editing were required. Because these subjects are rapidly changing, a substantial update was necessary. For this purpose, we first selected ten interviews out of thirty, and then conducted a series of back-up interviews, assisted by the editorial staff in the Tokyo office of Oxford University Press. After this new series of interviews was completed, we reorganized the material, integrating old and new interviews. Thus, the present book is not a translation of the Japanese edition, but a new book based on it. I would like to express my gratitude to those original interviewers whose questions have been somewhat 'transformed'. These are as follows: Chapter 1, Mr Etori of Mita Press; Chapter 2, H. Yanagida of the University of Tokyo; Chapter 3, Mr Etori; Chapter 4, Professor Murakami of the University of Tokyo; Chapter 5, Professor Karube of the University of Tokyo; Chapter 6, Mr Moritani, a journalist; Chapter 7, Professor Nakamura of Waseda University; Chapter 8, Mr Etori; Chapter 9, Professor Nakamura; Chapter 10, Mr Nano, a journalist. Also I appreciate the efforts of the translators and the staff of Oxford University Press for carrying out this difficult project.

OKOSHI

ONE

CASTING LIGHT ON THE PAST AND THE FUTURE

TAKANORI OKOSHI

OPTICAL COMMUNICATION, which utilizes light for signal transmission, is expected to replace the current electric telecommunication systems during the next century. It is essentially a post-war development, initiated in scattered laboratories around the world. Japan was one of the countries where this technology was first conceived, and also where the actual development of devices and systems is most actively pursued. Professor Okoshi is one of the leading experts on optical communication technology. After graduating from the University of Tokyo in 1960, he went to the USA to work in Bell Laboratories, where he witnessed the initial phase of the development of optical communication technology. After returning to Japan, he made important contributions to optical and imaging sciences. Among these, the most important was the proposal for and research into coherent optical communication, the most refined mode of optical signal transmission. He has been active not only in academic research on the technology but also in liaising with industries, the government, and the academic community to make optical communication a major field of technological advantage of Japan. He is now President of National Institute for Advanced Interdisciplinary Research (NAIR).

In this interview, he focuses on the historical origin of optical technology, his contribution to it, and his predictions for its future. He also analyses and criticizes the cultural environment of Japanese technology.

———

'I believe you were born in Tokyo.'
Yes, I was born in Bunkyo-ku, Tokyo.

'Was your father a scientist too?'
Yes, he was a professor in the Faculty of Engineering at Tokyo Imperial University.

'So, I suppose it was natural that you went into engineering?'

Well, yes, but we had a kind of family rule not to follow in the footsteps of our fathers, so my father didn't want me to go into mechanical engineering. Actually my grandfather was a graduate in electrical engineering, and my father had been told not to follow him either, so he chose precision engineering instead. Thus, I have three children none of whom went into electronics.

'Were you interested in engineering from childhood?'

Yes, I was quite interested in taking up science from elementary school up to junior-high. But when I was at high school, I thought of taking up a literary subject.

'Why was that?'

I became interested in philosophy at that time under the influence of friends and I read philosophers such as Kant, Schopenhauer, Nietzsche, and so on. I felt, though, that there wasn't much future in philosophy; it didn't seem a very positive discipline. But then I thought, rather flattering myself, of becoming a sort of critic of civilization. Because of this I wanted to go into the literature department, but that was just a fleeting dream, and thinking back, I'm glad I didn't go into that field.

'Was electronics a popular subject among students at that time in Japan?'

The most popular subject was theoretical physics. Electronics seemed to have been next and was quite popular. Applied chemistry was popular too.

'Were you fond of playing with electricity when you were little?'

I liked it just as much as anyone, and I remember making maybe four valve radio sets when I was in primary school.

'When you chose electronics, had you already decided to be a researcher?'

To be truthful, I thought about it, but wasn't sure if I could do it. I thought vaguely that if I were to become a researcher, I would do theoretical physics at first. Going into electronics was a sort of compromise.

'Did you go into electrical engineering because you could make a living in it whereas you couldn't in theoretical physics?'

Yes, that was one of the reasons. I wasn't confident of my abilities, so as you say, if I continued in electronics I could be close to physics and yet make a living as an engineer. I was sort of deferring a decision.

'Theoretical physics sounds quite basic, so whichever course you decided on, were you interested in basic research?'

Yes, I was. But I wasn't sure if this was realistic. I was about twenty; anyone, except those with great talent, might have felt the same.

'So at what time did you feel confident that you could go on as a researcher?'

That's quite a difficult question, but I suppose it was when I was thirty, when I joined Bell Laboratories. Excellent researchers gather there from all over the world, and when I found myself among them and after competing with them, I felt for the first time that I could do it.

'But was it a difficult decision, going to Bell Laboratories?'

It was mere luck. Not just anybody could go. And in any case the standard of research in Japan was quite low.

'Were you already interested in optical technology?'

I wasn't at the time, but something quite surprising was happening at Bell Laboratories when I joined in 1963. In Japan, people were still working on microwaves (with wavelengths of a few millimetres). The problems were, however, becoming apparent. With microwave technology, a technique was needed to make a circuit of comparable size to the wavelength. Thus, while doing research on millimetre and submillimetre waves, we wondered whether such a technology was practical.

'So, you were in a sort of state of anticipation?'

When I went to the US, there was, in my department, a Greek American researcher called Zacharias, and while I was there, he had to change his subject. I asked him what he was going to do next, and he told me he would specialize in the calibration of lenses. He explained that he had been asked to calibrate or systematize the checking of the precision of these tools for optical science for use at Bell Laboratories. To my surprise, his reasoning was that, since it was the era of optical technology, surely research would be focused on optics, and so he would be prepared for it.

'Did you come across other signs of the coming "era of optical technology" at Bell?'

Well, in my department research in holography had already begun. That was done by an English researcher called Pennington. Research in holography took off in 1962, a year before my visit to the US. It started at the University of Michigan, but a year later experiments in holography were initiated at Bell

Laboratories by Pennington and others. I had many discussions with him, arguing that Bell Laboratories should be a place for telecommunication research, and wondering why he was doing research in optics.

'So you were not convinced that it was the era of optical technology.'

Pennington's argument was that the time had come for optics. As you know, in holography images are created in three dimensions by recording phase information of light on to a photographic plate called a hologram, and then irradiating the plate. He told me that this had certain possibilities in information technology such as optical memory, and eventually I agreed. At that time, research in holography at Bell Laboratories was kept secret from the outside world. I suppose that was the end of 1963.

'So you recognized that a new era was dawning?'

Yes, I saw that the new field of optoelectronics was opening up between optics and electronics. At that time, there were a number of researchers working on lasers. Also, I remember a person called Loebner of RCA writing a famous paper on the importance of optoelectronics devices, in 1955.

'That was before the laser was discovered.'

Yes, so the recognition of the importance of optics, especially after the laser appeared in the early 1960s, had already been made at Bell, though it wasn't that general. It was that sort of time.

'Of course there was hardly any research in optoelectronics in Japan at that time, was there?'

Research on lasers had started gradually from around 1963. In Japan, at first, it was more common to say 'laser optics' than 'optoelectronics', or words like 'quantum electronics' or 'quantum engineering' were used instead. The word 'optoelectronics' was established after the mid 1960s.

'You were at Bell for a year and a half. Did that period mean a lot to you?'

Yes, it did. What meant most was that I was able to observe Bell's research management, though at that time there was a great difference, both economically and culturally, between the research environment of Japan and that of Bell, so there wasn't anything that could be adapted right away in Japan. Apart from that, I became more confident in my ability to get along with foreign researchers, and this meant a lot to me.

'So you maintained great confidence despite severe competition.'

Well if I dare to say so, that was the first time I felt confident.

'After you returned from the US to Japan, your salary dropped a good deal didn't it?'

Yes, quite a lot, but I thought it couldn't be helped. Moreover, I was thinking that someday, though I wasn't confident, we could bring the level of the Electronic Engineering Department of the University of Tokyo up to that of Bell.

'Did you feel that strongly after you saw what was happening at Bell?'

At a human level, I thought it could be done. When I compared the ability of young researchers at Bell with those of Tokyo University, I thought ours as good or even better. Of course there was a great difference in actual accomplishment at that time, but I thought my Japanese colleagues could shine if they were polished enough.

'It must have been difficult to feel that in the 1960s, when other Japanese of the time generally thought they couldn't possibly catch up with the US.'

I was rather stretching myself, I think.

'But to have that feeling must have been important.'

I did have the feeling that I didn't want to be defeated, though I really appreciated my colleagues at Bell, and was especially grateful to my boss at that time, Dr James W. Gewartowski.

'After you returned to Japan, you brought out a unique device called a soft-landing collector. When was it that you seriously got into optics?'

Various research projects were crossing over. It was in 1967, that I moved in to optical technology from three-dimensional imaging.

'Could you say more about three-dimensional imaging?'

My father-in-law was a physicist at the Metal Research Institute at Tohoku University. When he returned from a trip to the US during a summer vacation, he gave me a stereoscopic post card as a souvenir. Nowadays no one would be surprised at it, but at that time I was quite astonished, and thought if I combined this technique with that of holography, it would lead us to a three-dimensional television. I had been interested in three-dimensional imaging since the discussion with Pennington on holography, and now I thought of going into optics working on three-dimensional imaging. Furthermore, I had discussed the importance of optoelectronics already in some committees.

'Optics is rather an old subject in physics.'

Yes, it is a traditional subject and there were many discussions as to how it would develop. Progress in optics was very slow and minutely incremental; for any real innovation to occur electronics had to be brought in.

'So Japan had now arrived at the era of optoelectronics?'

I remember at the end of the 1960s, we discussed such things at a symposium of the Japanese Society of Applied Physics. There were occasions after that when optics and electronics meetings collaborated, but in Japan cooperation between researchers within these two fields was not as extensive as that in the US.

'However, the level of optoelectronics in Japan has recently advanced, and is even said to have overtaken the US.'

That is what the Americans like to say.

'Back to three-dimensional imaging research, how did it get started?'

I started research on three-dimensional imaging in 1967, and in 1980, I wrote an invited paper entitled 'Three-dimensional displays' for the Proceedings of IEEE. This tutorial review paper was a result of my first intuition that the combination of lens-sheet technology (as in my father-in-law's post card) and holography might have a future. As a matter of fact, that was the last thing I did before I gave up researching three-dimensional imaging in around 1980.

'Why?'

Not for any special reason; I felt that I had done my best for the time being. I thought that I couldn't get any further without better resources. Another reason was that, a year earlier than that, I began to devote my attention to research on coherent optical communications, and so, because of lack of time, I gave up three-dimensional imaging after working on it for about twelve years.

'Now that you have moved into optical communication, can you tell me about the coherent optical communications that you have been working on?'

Coherent optical communication is a telecommunication system using phase-consistent light. That is, while in ordinary monochromatic light the light waves have various frequency components, a coherent wave has only one frequency.

'When did you get into research into coherent optical communications using these waves?'

I think it was in the early summer of 1978, though I had a vague idea about it a year or two earlier than that.

'Was there some "inspiration"?'

I had been giving lectures on basic electromagnetic waves to undergraduates during the winter terms, and in each opening lecture, I had described the classification of electromagnetic waves. From the lower frequency, that is from the longer wavelength, it goes very long wave, long wave, medium wave, short wave, ultrashort wave, up to supershort wave, and finally millimetre wave and submillimetre wave. The present custom among engineers up to the submillimetre wave, is to categorize by frequencies such as for example from 3 gigahertz to 30 gigahertz (Super High Frequency), or from 30 gigahertz to 300 gigahertz (Extremely High Frequency). However, from around submillimetre wavelengths, where infra-red rays start, waves suddenly get categorized by wavelength rather than frequency. During the lecture I explained it in the following way. In the technology of radio waves, the size of antenna is determined by wavelength and this is important. However, much more important is the frequency, as this determines how much information can be carried by the wave. Also, in the detection method now used in heterodyne radio receivers, the radio signal is detected by making a 'beat' with an electric oscillation generated in the radio which has a frequency close to the wave received. Hence frequency conversion becomes important.

'Then why is wavelength used in light?'

That is because, at least until recently, there was no application for light in which its frequency was critical. Rather, gratings or thin-fibre filters which essentially need the size of wavelength, were the main technology, so I taught that it was more convenient to characterize it by wavelength.

'But I think you did not altogether completely believe what you taught.'

Thinking it over, I realized that the dials of radios I saw in my childhood were marked with wavelength! Some ten years afterwards those radios from the mid 1940s until the 1950s mainly had frequency scales, but sometimes they also had wavelength scales. The period when wavelength changed to frequency corresponds to the time when heterodyne replaced the four-tube-direct detection radio receiver that was common before and during World War II. Heterodyne radios work by frequency conversion, so it is inconvenient to use wavelength.

'So that's what inspired you to look at light in terms of frequency and not wavelength?'

Reflecting on this, it occurred to me that some day even light may be measured by frequency and not wavelength. This idea took on substance in the summer of 1978. I discussed it at a symposium of the Federation of Japanese Electronics Societies entitled 'Wavelength multiplex optical fibre communications' in Autumn 1978, after I had brought up the question in the paper sent to the society. I remember predicting that though now we talked about wavelength multiplexing, in the future it would be called frequency multiplexing.

'What reaction did you get?'

Everyone fell silent and I was quite embarrassed.

'Did you then put forward the idea of coherent optical communication, too?'

I was wondering at that time if we were to learn from the history of the electromagnetic wave, it might well happen that optical communication would become 'heterodyne', though I didn't go that far at the time. In any case, it was then that I began looking into the possibility of coherent optical communication, and I talked about it twice during the following January and February of 1979. This turned out to be my first formal proposal.

'So your idea came from quite an elementary question.'

I feel that the questions teachers come across during lectures, especially in basic lectures for undergraduates, have great possibilities for creating good research subjects. It is true that more questions arise during graduates' lectures, but these are often rather trivial. This is the first time that I talked about my own experience. I often suggest to young teachers, as if joking, that they may gain something if they prepare lectures very seriously, without telling them about my own experience.

'I would understand from what you have said that coherent optical communications is in the mainstream of future research in optical communication.'

Yes, a very large number of papers on it have appeared.

'What's more, Japan, it seems, is the leader in this research.'

I presented the first paper on it in 1979. Then came France, the UK, and Germany.

'Has it been difficult leading the field?'

To be a pioneer is very hard in a way, but also easy, for you don't have to read enormous numbers of papers. If you are following somebody else, you have to read a lot in order to speak out, and for this you need great energy. If you are the front runner, you only have to read the latest publications and it is easy to exchange opinions at conferences, where you are able to receive most of the information. Thus, at least in this field, the fact that Japan was ahead even if only for a year means a great deal. We presented our paper in 1979, the French in 1980, the British in 1981, and the Germans in 1982. The Americans came quite late.

'Is coherent optical communication almost established both theoretically and practically?'

Yes, it is. Now it's a matter of economics and this is not easy, because it is a sophisticated technology. But working examples have already been established, though difficulties will continue for a few years. I suppose a time will come in the somewhat distant future, when this technology will be widely applied. The choice among the technical possibilities is like that between direct-detection and heterodyne radio; it is not technically easy. First we need a good laser, both tunable and yet with a pure spectrum.

'How do you feel about the future?'

I think it will be all right.

'Will it be a semiconductor laser?'

Yes, that was one of my concerns when I started research in 1978. When I proposed establishing a coherent system using semiconductor lasers at that time, there must have been people who considered the idea crazy. From the point of view of common sense, the spectrum of purity semiconductor lasers seemed awful. However 1978 was the year when others discussed the possibility that the spectrum of a semiconductor laser could be improved much further if the spectrum purity of the predicted theoretical limit could be reached.

'How did the research proceed then?'

There had already been some reports at that time that the spectrum purity of semiconductor lasers had become much better for unexplainable reasons. These reasons gradually became clear: it was because of reflection of light from far away. When there is a reflection from a certain distance, for example from fifty centimetres, the laser works as if the length of the resonator is the same as that distance. Then a very sharp spectrum appears. Knowing this, no matter how mysterious the reason, I remember thinking that, in the future, coherent optical communications could be achieved with semiconductor lasers.

'So it was already possible to make semiconductor lasers with a precise frequency?'

Yes, what we did at that time was to first stabilize the central frequency. I started work on coherent optical communication in the summer of 1978, and a year later, we stabilized the frequency for the first time. The fluctuation per hour soon went down to 10 megahertz. At the same time, since there was no way of measuring spectrum line-width, we invented the 'delayed self-heterodyne method'. As a result of measuring with it for about a year, we found that there were cases where line-width could be reduced to 20 mega-hertz. This was reported in a paper which is often referred to. This 'delayed self-heterodyne system' has been made operational by Japanese companies and is already in use all over the world, though I don't have any patent for this and haven't received anything for it.

'What was contained in your 1980 paper?'

In that paper I described two advantages of coherent optical communication. One is, because of improved sensitivity, the repeater distance could be extended to about a hundred kilometres. Secondly, as used in radio, high-density fre-quency-multiplexing could be realized. I claimed that frequency fluctuation would be reduced greatly if we worked hard on the application of semiconduc-tor lasers. This was the first paper on coherent optical fibre communications.

'What was the reaction?'

In the following year, at the largest international conference yet on optical communication which took place in San Francisco, I gave a lecture as an invited speaker. Those who were in the front row were the leaders of the French and British Laboratories and I was asked a lot of difficult questions. Thinking back now, I realize that they were taking it seriously; though the US didn't show any sign of interest yet.

'That seems strange.'

There was a dispute in the US; young people thought they would be left behind if they did not take on coherent optical communication, but others were more cautious for economic reasons. That may have been a good way of thinking, but also, among the leaders there was a good deal of questioning about the weakness of the current optical communication, for example that it used a noisy optical carrier, and was of very low quality.

'Even using a laser?'

Yes.

'And they didn't go for coherent optical communication?'

No, but I suppose in the US most of the discussion was about what was wrong with it, because a 'low quality' approach leads to low cost. In November 1983, I went to Bell Laboratories and gave a lecture. A room holding about fifty people was full and almost all the famous people in optical communications and other fields at Bell were present. I talked about the research situation and asked them why they wouldn't go into it because they really should. I went to Murray Hill Laboratories the following day, where I again said that Bell would be left behind if they stayed as they were.

'Was that the first time you went back to Bell?'

I returned to the US about seven or eight times after leaving Bell, but at that time, I was cautious about speaking out. When I did, they listened with smiles on their faces. Internal dispute and the views of their leaders prevented them from speaking up. Thus I was the first person at Bell to advocate coherent optical communication. After that, they caught up straightaway. Now, major research organizations include two Bell Laboratories, NTT, and laboratories in Britain and France. Then come NEC, KDD, and Fujitsu although they are on a smaller scale.

'Are those laboratories all competing with one another?'

Because they have to undertake experiments at high cost, in terms of money and people, not all of them do. NTT, ATT Labs, and Bellcore have the highest level of research, probably followed by the British, Fujitsu, NEC, and KDD. France is doing fairly well.

'So you should be credited as the first person to propose the idea of coherent light communication?'

Yes, but anyone could have anticipated the progression from optical fibre to coherent light communication; the really important factors are the various small ideas in between. You cannot just say 'make it coherent!', for there are lots of technical obstacles in the way. It's no use if one cannot come up with ideas to overcome those obstacles.

'Are these successfully overcome in those laboratories you mentioned?'

Well, I would say yes. However, the disadvantage of huge laboratories full of people and money is that in spite of their scale, they generate few ideas, as they tend to get bogged down in large-scale experiments. Since the research into coherent optical communications began, the contribution of ideas from universities has been relatively high, although the teamwork has been weak. On the other hand, universities cannot try large-scale experiments owing to lack of personnel and money.

'What seems to be a bit strange is that no American universities are included in your list.'

They hardly deal with optical communication research. There are some who do, for example at MIT or Princeton. That's the difference from Japan where, in a sense, the distribution of researchers is wider. This could be an interesting theme in the theory of research management. In the US, not just optical communication but communication research generally is concentrated too much at Bell. When I chat with people from Bell, a few admit that what I am doing at my university is what Bell should be doing.

'So would you say Japan is superior to the US on average?'

According to a survey in the US, there are two fields in which the level of Japanese research is higher: one is in production technology, and the other is optoelectronics. Perhaps this has something to do with the fact that in Japan researchers are scattered widely between universities and industry.

'That's an interesting theory.'

I'm not sure about that since I have just come up with it now.

'It seems, considering the general research and development situation at this point in Japan and the US, the respective levels are reversed as far as optical communications is concerned.'

In Japan, there are many really good researchers, especially those who lead with new ideas at universities, though these universities are weak when asked to take on anything on a large scale.

'What do you think of the attitude to research in the US?'

I suppose Japan can already compete with the US in quality, but the US has more than twice the quantity of researchers. However, I remember being asked when I returned from Bell, how such a luxurious system could last forever. In the US there has always been a tendency towards 'big' projects, without regard for the cost, and there is the problem now that financial difficulties have become apparent. Also, as a basic cultural matter, there seems to be too much emphasis on brain work disassociated from 'physical labour', the thought being that a researcher should not carry anything heavier than a pen.

'Do you think the attitude of Japanese research workers is more balanced?'

Yes, in a sense. Often, engineering researchers work together with technicians.

'In what way do you think the range of coherent optical communications will grow as the twenty-first century approaches? It seems certain that eventually it will be central to optical communication technology.'

It will take some time before this happens. In the next century, the day will come when a great proportion of optical communications will be coherent; just as radio has become heterodyne and coherent, so will the optical technology. However, before we get to that stage, we will have to overcome many technological problems to make it work at an economic level.

'Although there are still financial and technical constraints, if coherent optical communication does take off, it will be able to receive all this minute information without mixing it up, right?'

One example would be a high definition television (HD-CATV) network, though this is not yet practical from a cost point of view.

'At first, the spread of HDTV seemed to have been for specialist use rather than for television broadcasting itself, but if HD-CATV comes into use along with coherent optical communications, it will be ideal for broadcasting. So, it is certain that optical technology will become one of the key factors in the technology of the future.'

It is a matter of technological leadership in the twenty-first century. Japan and the US have been strenuously competing over this; the US especially has been pushing the research through under government leadership.

'If optical communications spread into various fields, will electronics become of minor importance?'

I'm not sure about that; I would take a conservative attitude. Probably the near future will not be all optical because electronics does have merit. I don't see any reason to only look for optical systems.

'So you think it's better if optics coexist with electronics?'

Yes, definitely, as far as the near future is concerned. After that, I'm not sure what will happen, for instance fifty years on. Maybe for the next twenty years.

'I'm sure industry sees a huge market in optics.'

Yes, it is said that optoelectronics will be a twenty trillion yen business in the year 2000. Current electronics is around nineteen trillion, so it would be comparable. I should say that by then, there may not be a division between them. Even now, it's quite difficult to draw a line.

'The size of the market grew enormously in the eighties, I believe.'

It's been said that there's hardly any comparable field that has grown ten times in ten years; it must be a record. By the end of the eighties, it had grown about a hundred times in ten years. That's about the rate shown by the semiconductor business at one time, a yearly rate of twenty-six per cent.

'Will this increase cause friction with other countries? How could this be avoided?'

To give an example, video cameras and compact disc players are monopolized by Japanese industries; this is not the case with optical fibres. It isn't a problem yet with optoelectronics, but the possibility does exist.

'It would be a question of whether other countries can produce optical devices. If not, then Japan has to have the monopoly. Do you think that would be right?'

The advantage Japan has in large-scale integrated circuit manufacturing is due to various factors such as a relatively high educational level, superior labour work, the willingness of the young, and active funding of enterprises; it would take any other country something like ten years to catch up. However, companies have set up factories and laboratories abroad to avoid such a monopoly. I really think Japanese companies are making an effort. However, Japan is still weak in new ideas; for instance the original compact disc was made by Phillips, and television was invented in the US. Even the basic theory of optical communications was developed in the US, but Japan quickly caught up with it.

'Do you think then that Japanese researchers should do more basic research?'

The question of whether Japan is taking a free ride and has no originality has now become a stereotype. This question differs in each field. Also, in the final analysis, it is a matter of the strengths of the individual researcher. I myself like to do basic and systematically structured work, and prefer to study the history of a subject rather than merely follow fashion. But this is a matter of personal preference and not national policy. Although a successful country sets out imitating others, there is a tendency for it to forge ahead of the pioneer countries who have become inert through complacency. Japan's originality will bloom in the future.

'So you are optimistic about our creative future.'

I don't see any definite reason why there should be a call for basic research in a nation. It has, after all, to do with the nation's sense of its own dignity.

'Do you mean that dignity is now an important factor in Japan?'

I would like to think so, but I'm not sure it is time yet. Japan will have to use its technology to further social development. In this sense, I think Japan is still a poor and backward country.

'Do you think we ought to be even richer?'

Yes, in a sense I do. I still think that in Japan technology is not yet wholly serving society.

'That's an important point. It is true that it serves the nation in a sense, but not so much individuals. What should we do about this?'

I suppose it's a matter of choice which values are adopted. Let me give one small example. For the last ten years I have been advocating that we bury electric and telephone cables underground; and I'm even thinking of taking up this issue as part of my life work. Japan is the only one among the many developed countries that have electric and telephone cables and poles in the midst of its towns. However, we have the most advanced optical and electrical communications technology in the world, which seems a contradictory position.

'I think, though, this is a complex problem.'

The problem of burying electric cables underground is quite interesting. As a matter of fact, if we study this, a chart of the various conflicts within Japan's technology, administration, and society might be drawn up. Despite its abundant technology, Japan does not have the ability to use it to build beautiful towns the way other advanced countries are able to do. One of the reasons for this is that we are unable to obtain the right balance between these various interests.

'In essence this is a problem of urban design?'

I would say that most of Japan's cities are slums. Although we have world famous and much-praised architects like Kenzo Tange or Noriaki Kurokawa, they have done little to solve Japan's slum situation, which is a pity. But I'm not blaming them; I couldn't have done anything either, even if I were Mr Tange. What's interesting about the post-war world around 1950 for instance, was that since a great deal had been destroyed, we could start again from scratch. Then everything gradually settled down, and within ten or twenty years, the outline of society had been drawn. After fifty-five years, for good or ill, each element in society has been defined. This has its merits, but on the other hand, it has led to excessive specialization causing difficulty of communication at all administrative levels.

'What's the government's attitude towards this sort of opinion?'
First, I would like to say that the government itself is quite rigid in its attitudes towards communication. The example I gave of electric cables shows the government's inefficiency, its 'underdeveloped country' mentality. As for the relationship between researchers like us and the government, I would say the government tries to take advantage of us; even though, from abroad, it might look as though we are actually influencing the government. The attitude of the Japanese government is very much to take advantage of what others have created.

'Have you ever considered whether there are limits to technology?'
Well, for example, the telephone was welcomed by society a hundred years ago. Will this be the case with a video telephone? It wouldn't be good to go too far. Even now, electrical appliances equipped with so many controls have become too difficult even for professors of electrical engineering!

'I hear one of your hobbies is music.'
Yes, I like listening to music, and I am even an amateur composer.

'What sort of music do you compose?'
I would say it is unashamedly classical; I only listen to classical music. The first concert of my music was performed in 1989. It began with Chopin and then there were ten of my pieces. I consider this to be quite an honour.

'Was it orchestral?'
No, it was scored for flute and piano and played by first-rank professional musicians.

'So you're really only into classical music.'
I suppose I am a man of quite extreme opinions; I think within fifty years, modern music will have mostly disappeared.

'Then what do you think will be left?'
In Japan, I would say some vocal music will survive. For example, the songs of Yoshinao Nakata will remain, but as for the rest, such as the 'ultra-

modern' pieces composed by the so called 'avant-garde' composers which are now winning prizes, these will be forgotten. I don't think this is only my opinion. I don't know about the professional music critics, but there are some ordinary music lovers like myself who share my views.

'So do you think modern music has become too technological?'
Well, 'ultra-modern' composers seem to go too far in seeking new ideas, for instance rejecting tonality. Human feeling cannot keep up with it. Even now we listen to Mozart with pleasure, don't we? I do listen to music which won first prizes in the 1980s just from intellectual curiosity, but there is not much of it that I would like to listen to over and over again.

'Have you ever had any lessons?'
I'm only an amateur; I am self-taught. I had lessons from two piano teachers, but was sent away by both because I brought them lots of trivial pieces I had composed!

'Then you play the piano quite a lot yourself.'
I should say I must be the worst pianist in Japan.

'Changing the subject, whether it is related to your studies or not, what would you like to do in the future assuming you have the time?'
First of all, I would like to continue my research into optical communications. I recently published a short paper on the theory of optical amplification. Also, since as I get older I get more opportunities for public speaking, I hope that I will be able to contribute a little to improving the links with other fields. I would like to help make Japanese cities beautiful, by solving the problem of burying electric and telephone cables underground. I think this is a responsibility for us engineers.

'Can I ask you what your private ambitions are?'
Well, my family is originally from Tochigi prefecture, some 100 kilometres north of Tokyo, but since nobody has lived there since my grandfather's time, I would like to do research on it and write a family history. The register of the dead was burnt in a fire ninety years ago, so I can't go beyond that, except from the inscriptions on tombstones. I can only study in detail the period since my great grandfather's time; he died in Tochigi.

'What did he do?'
It is not clear because there are no survivors from the country. I'd like to go and look around in the cellar and record the details on the tombstone. The

last three generations of the family including myself have all been engineers, but no one had been interested in doing these things. Since the country is unchanged, I would like to sort out these facts before I die.

SUEMATSU

TWO

CREATING A TECHNOLOGY
AND TECHNOLOGISTS

YASUHARU SUEMATSU

OPTICAL COMMUNICATION systems have three main components: the source, the guide, and the receiver of light. The best light source at present is the semiconductor laser, which was actually the synthesis of two post-war key technologies, semiconductors and lasers.

Professor Suematsu is one of the leading researchers into this synthesis. He graduated from the Tokyo Institute of Technology (TIT), the institution for fostering technological cadres in Japan, and has been engaged in researching the manufacture of semiconductor lasers for use in optical communications. His research during the 1960s and 70s is representative of Japan's advancement to a state of leadership in optical technology throughout the world. He has also been an outstanding leader and teacher within TIT, of which he is a former President.

In this interview, he starts by describing the development of optical technology from his own personal viewpoint, and then discusses the status of technology and technologists in Japan. Finally he talks freely about the cultural environment of Japanese technology.

———

'Japan is now one of the most advanced countries working in optoelectronics technology, both in industry and in research. Why do you think this is?'

One reason is accidental. Optoelectronics took off around 1960. At that time, Japan was still in the midst of reconstruction after the war, both of its industry and its universities. Technologists working there were innovative and ambitious. Another reason is to do with the academic tradition in Japan. Japan was heavily involved in research on microwaves. Microwave technology was already being applied in communications. Reducing the generation wavelength of the millimetre wave was proving to be a problem. It was actually my subject at graduate college. The concept of optical communication seemed somehow a natural solution to this problem.

'At that very time, I believe, you had started lecturing at the Tokyo Institute of Technology?'

That's right. At about that time I decided to go into research to explore the possibilities of optical communications. While I was still an undergraduate, Professor Towns,[1] a Professor at MIT, came to Japan and I attended his lectures on masers. In my graduate years, the limitations of the electron tube for short millimetre wave generation were becoming apparent at a time when the future was beginning to look better from the socio-economic point of view. I had the strong impression that we would soon see the light at the end of the tunnel.

'How did you come to focus on semiconductor lasers?'

Optical systems have three basic components: a light source which emits the signal; a transmission medium which transmits the light; and a light receiver which receives the signal. As far as the light receiver was concerned, we already had electron tubes. The photo-diode came relatively early too, so there would have been much of a problem if it hadn't been for the noise. Noise levels depend on the quality of the crystal and its surface processing, and the facilities we had at the university were quite inadequate to deal with this.

'You couldn't achieve uniform quality?'

No we couldn't, so I decided to work on the light source and transmission medium instead. Two or three years later, one of my students told me that something called a laser had been invented and because it looked very interesting he wanted to organize an exhibition for the Institute's foundation day. I pointed out that there was no glory in merely borrowing lasers from the manufacturers and I suggested that he build his own optical communications system. That was in 1963, three years after the laser was first perfected. To modulate the light, you need a KDP [potassium dihydrogen phosphate] crystal. This crystal was made by Professors Ogawa and Namba of the Riken Institute.

'What did you use for the light source?'

I used a gas laser. The modulated light was to be received by a photomultiplier, so we applied a voltage to the KDP crystal to modulate it and as a result actually made a few experiments in optical communications. At that time, quite by chance, Canon announced that they had developed a new fibre for endoscopes. We decided to try passing a signal through this fibre, and found that the noise level was surprisingly low. As a result we used the KDP semiconductor lasers for communications. But before we rushed into single mode, we needed a theory to prove it was really so. I discovered the

---

[1] The inventor of the maser.

theory based on electron relaxation around 1970, which laid a firm foundation for single mode, but it was ignored at the time.

'I think also you attacked the problem of optical medium at that time.'

Yes, we did. As I said earlier, using the optical fibre we had at the time, the signal fell to a fraction of its initial power over as little as fifty centimetres. Over a hundred metres of fibre, the light would disappear completely. At the time, a number of people, including Professor Kao of STL [Standard Telecommunication Laboratories] in England, were working on practical methods of improving fibres so as to avoid this loss. I was asking the same questions, but nobody had any answers. We experimented with a completely different method, in which light was passed through a series of lenses, and we found that if the lenses were lined up carefully over a distance of a hundred metres, the loss was very small. However, if the individual lenses were moved by so much as a hundred microns, the light was lost over a matter of a few kilometres. We gradually became aware of the phenomenon of beam instability.

About 1968, Japan Glass Ltd and NEC, and Dr Kao as well, were looking at ways of reducing loss in optical fibre. In 1970, Corning developed a fibre which had an unbelievably low loss of only twenty decibels per kilometre. So industrial organizations had begun work on optical communications by improving fibres to reduce loss.

'So in the seventies, the gap in technical achievement between America and Japan vanished. How were you able to make such rapid progress?'

Well, it's this kind of practical commercialization that brings out the originality in Japanese people. Whenever a new technology is developed, there are always people who come forward and say that they thought of it a long time ago. But that means nothing. To actually realize something, to develop a practical application, is very difficult. And that's what Japanese engineers are best at. There is one more interesting point—the relationship between scientific communities and industry in Japan. Let me give you an example which I feel is particularly striking. I belong to a professional society which meets every month. Many of the meetings are held in the reception rooms of industrial companies. That kind of informal intimate cooperation, I think, is quite uncommon in the West.

'I see. One of the best examples of the progress made during the seventies is your development of the integrated semiconductor laser. Could you tell us a little about it?'

I was thinking about integrated circuits of optics when I began my research on lasers. As there were integrated circuits for microwaves, I thought I must then be able to integrate optical elements. One of the basic reasons I decided to try my hand at making a semiconductor laser was that integration was possible given the fact that a semiconductor laser is formed by an optical waveguide. In 1965, I wrote a paper on the waveguide theory of semiconductor lasers. I speculated that it would be possible to integrate an optical waveguide circuit—for example, a guided-wave parametric device such as a tunable parametric oscillator circuit. And sure enough, in 1969, one of my research assistants found that the latest issue of a British journal contained an article on integrated optical circuits. I said, 'Oh, so they have thought of it too, have they?' and we were not at all surprised. Research at that time was moving in the direction of the integration of optical circuits anyway.

So, I wanted to work on integrated optical circuits for semiconductor lasers, but I had only just been made an associate professor and my research funding was not generous.

'So, even you, a leading technologist in Japan, had a hard time getting money for your research!'

At the beginning of my research career, as I have said before, I was in a new field, I hadn't even got an optical bench of my own! So I had to borrow one from the physics department. Though I had some support from industry, I was certainly in a difficult position, trying to build my own semiconductor laser. In 1967, I managed to make one using a gallium arsenide p-n junction, but then I ran out of money and couldn't go any further. I really wanted to build semiconductor-integrated optical circuits, but to do that I needed an electric furnace. At that time, thanks to Professor Suetake, the engineering faculty offered to lend us three million yen. That was a lot of money in those days. We wondered how we would ever be able to pay it back, but they gave us three years, so we were able to buy our electric furnace for crystal growth. That was in 1973. Anyway, we began work building an integrated laser. To do this you need a waveguide circuit with low loss and, because the part which generates the light has high loss, you have to find a good way to connect it to the waveguide circuit. We hit upon the idea of stacking two waveguide circuits, one on top of the other, to make a directional coupler. This became the 'integrated twin-guide laser'. It was completed in the summer of 1974, and was officially the first laser of its kind. Several months later, Texas Instruments and Bell Laboratories came up with their own versions.

'What was the situation with the optical fibres?'

At that time, the quality of optic fibres was gradually improving and, thanks to great efforts on the part of industry, Japan was producing fairly good lasers. However, research by NTT showed that when a signal was modulated for transmission by laser, the transmission characteristics deteriorated. You see, with the semiconductor lasers we had at the time, when you modulated the light, the spectrum broadened due to multi-mode operation. The integrated laser that we proposed controlled wavelength so that, even after modulation, the spectrum kept a single frequency—in other words, single-modal. That was around 1977.

I asked Dr Iga, who was an associate professor at the time, to make some calculations on our proposed integrated semiconductor laser, to work out whether the spectrum would remain monomodal under dynamic, high-speed modulation and how far it would be possible to suppress the lateral mode. By and large, he found that we seemed to be on the right track, so we decided to try light modulated in even shorter pulses, with a width of the order of one nanosecond, in other words about one gigabit, and see if the spectrum stayed single-modal. An integrated laser capable of emitting a nanosecond pulse at a single wavelength—a 'dynamic single mode laser'—had been perfected in 1979, and we used that. Sure enough, with pulses of 1.5 nanoseconds at any point on the waveform, the spectrum was a single frequency. So we had our light source.

'What semiconductor material did you use?'

To tell the truth, we arrived at the material by a rather roundabout route. Until around 1975 we were using gallium arsenide. At first, the focus of research was on optical fibres with a wavelength of 0.85 microns, but in a paper in 1973, Corning suggested that the ultimate loss of silica fibre could be reduced at longer wavelengths of 1.3 or even 1.5 microns. We believed the future was in integrated optical lasers in which the wavelength would be optimized for the optic fibre. So we decided to develop a semiconductor laser with a wavelength of 1.5 microns.

'So if I am right, you also developed the laser itself?'

Yes, we did. We started off with an indium phosphate substrate, and in 1977 managed to obtain a wavelength of 1.3 microns.

'It was quite an achievement to start with the raw material and get as far as you did.'

We realized in 1978 that it was possible to achieve a wavelength of 1.5 microns, and by 1977 we had succeeded in making the so-called 'long-wavelength laser' operate continuously. Meanwhile, as predicted, optic fibre loss was reduced. Before that time, research on fibre loss reduction and the development of long-wavelength lasers had been conducted separately, but from 1979 they went hand in hand. That year, NTT demonstrated that a fibre with an optimum wavelength of 1.5 microns gave a loss per kilometre of 0.20 decibels, and at the same time, a laser with a wavelength of 1.5

microns was at last achieved. People working on fibres like to say that their success paved the way for our achievement with lasers. I feel however that we were able to achieve a wavelength of 1.5 microns because we had predicted that optical loss would come down because of the very characteristics of fibres. In fact, we were working in parallel, and each development took about four years.

'You tried integration, didn't you?'

Yes, now that we knew that lasers with a 1.5 micron wavelength represented the way ahead, we decided to work on integration, achieving a single frequency spectrum and controlling wavelength in 1.5 micron lasers. A 1.3 micron laser had been perfected two years earlier, so for the time being we built a DBR Distributed (Bragg) Reflector-type integrated twin-guided laser and used it to show that a single-frequency spectrum was possible. As I have said, that was in 1979. In 1980 we proved that it was possible to achieve a single-frequency spectrum with a 1.5 micron laser. We published our findings at the International Conference on Semiconductor Lasers that year. We reported that we had managed to build only ten lasers of this type; because our work was still at the research stage, the threshold current density level was some ten times higher than that of the normal laser at that time, but at any rate, our laser had produced a single wavelength. It seems that nobody really understood the significance of having a single mode operation under rapidly modulated conditions, and in the eyes of most people, I had merely built an unusual laser with a threshold level ten times higher than that of most lasers. So I was given a prize for being the man who had built the laser with the world's highest oscillation threshold level! That seems typical of the British sense of humour.

'But in the eighties the single-mode laser gained strong support.'

Once the theory had been worked out in detail, it was clear that what suppressed the neighbouring longitudinal sub-mode was very weak in normal, or so-called Fabry-Perot lasers, so that even a small disturbance would be enough to generate another longitudinal mode. We realized that the conventional fabrication methods for the Fabry-Perot laser were not suitable for building a dynamic single-mode laser, and that we would need integrated lasers with a mechanism that selected wavelength structurally. So when NTT, which has the most powerful influence in this field in Japan, decided to go with monomode fibres in 1980, the dynamic single-mode laser became the focus of attention.

'You seem to have accurately predicted the future.'

Ultimately, it is the narrowness of the spectrum which determines the transmission bandwidth of a single-mode fibre. So everyone soon realized that dynamic single-mode lasers were the way ahead and they began to make them too. But industry carried on using 1.3 micron fibres. That was because those fibres which have zero dispersion could be used even if the light source spectrum was wider, as is the case with conventional lasers. But in around 1988 or 1989, a 1.5 micron system was developed and it is now certain that undersea cables in the future will have a wavelength of 1.5 microns. This is quite natural given that 1.5 micron fibres give reduced loss. It is illogical to ignore bandwidths with low loss and use fibres in bandwidths with high loss.

'What does all this mean for optical communication in general?'

We need very precise tuning of the laser wavelength in applications such as coherent optical communications. In order to do this, you need to tune the refractive index of the waveguide circuit from outside. The best we could manage using an applied field was 0.05 per cent. But we discovered that when a current was applied the refractive index changed by around 0.5 per cent. So one of my colleagues took the lead by building an integrated laser which incorporated a mechanism that controlled the refractive index by applying a current. In 1984 we built the first laser in which the wavelength could be tuned by up to 10 Å. We called it the 'wavelength tunable laser'. This laser evolved and is now very widely used in industry.

'How efficient are lasers today?'

The further you go with them, the harder it is to improve efficiency. At 1.5 microns, we managed around 30 per cent of the quantum differential efficiency. That was the greatest efficiency we could achieve with the given laser, and was not good enough. In the end, to improve efficiency we had to reduce loss in the laser active medium. Unfortunately, conventional bulky materials were not at all adequate. Then along came artificial materials such as quantum structures, and we began to feel that if we could construct quantum wires and quantum boxes, our problems might be solved. Theoretically, the amplification gain of the quantum box was twenty times higher than that of its bulk. We estimated that this would not result in a considerable increase of loss and therefore give us extremely efficient lasers, so we started work on quantum wires and quantum boxes. The first successful oscillation with a strained quantum box laser was achieved by our group in July 1993.

'The quantum box is an area of optoelectronics in which Japan is particularly advanced, isn't it? What is the position now?'

Well, it is really Professor Sakaki's field, but I have a feeling that we are now at a stage where we could build a quantum box if we really tried. But how should we go about it? Many methods, such as Professor Sakaki's slanted substrate method can be used for quantum wires. There is of course a need for the sort of ultraprecision engineering at which Japan excels. Recently—and this is the latest thing—scientists in Illinois have proposed a method which involves annealing from quantum film and which does away with the need for any processing whatsoever. How they came up with such original ideas is beyond the limits of the imagination!

'You don't know what to expect next!'

That's right, it's anybody's guess.

'Professor Suematsu, you are a pioneer in the development of optoelectronics, which is one of today's most important technologies, and you are also President of the Tokyo Institute of Technology, which produces excellent students as well as some of Japan's leading technologists. May I ask what are your general views on technology?'

Well, fundamentally, I believe that technology has become the basis of civilization. In the West, especially Europe, technology is little more than an occupation and a rather vulgar one at that in most people's eyes.

'At an unconscious level they regard Christianity as the foundation of civilization, don't they?'

Yes, but it may surprise many westerners that, even in Japan, technology is by no means regarded as the basis of civilization.

'Yes, the idea of technology as part of culture is not commonly accepted here.'

Although it may sound odd to foreigners, this is true; engineers are simply seen as people who beaver away making things. The leading figures in our society are lawyers and economists who see technology only as another way to make money. Just the other day, for instance, the head of a securities firm said that it didn't matter if Japanese technology worked out provided Japan keeps control of finance and information. I was shocked.

'So you think engineers may have to become philosophers?'

We will have to make sure that technology ranks at least as high as science.

'Yes, technology is not just an application of science.'

Well, take the things I have worked on all my life. For instance, the various devices in optical communication systems. The principles for their operation are found in the physics of light, the physics of materials, the mathematics of wave motion, and in information theory. But taken on its own, the knowledge of none of these sciences is enough to build such a device. To create what is required, you have to try to imagine an unknown structure with the help of a combination of known physical phenomena. This is device engineering. It is quite different from 'science' where you try to systematically understand existing phenomena in ever greater depth.

'In the West though, people tend to rank science above technology, don't they?'

Yes, especially in England, people rank science on a level with theology, philosophy, and mathematics. And engineering comes below these disciplines. In America, technology is highly regarded, but is treated very pragmatically and, as far as I know, nobody has tried to develop there a 'philosophy of technology'.

'Japanese universities were the first to establish departments of engineering in recent times. Engineering has not had the same standing in universities in the West!'

I believe the philosophy of technology is something Japan's engineers will have to tackle in the future.

'Can you give me your views on the education of engineers at the Tokyo Institute of Technology?'

Well, I try to emphasize that there should be a good balance between creativity and practicality. Actually, our university motto is 'Advance towards the intellectual–industrial frontier'.

'What do you think generally of the education of engineers in Japan?'

In the eighteenth and nineteenth centuries, England produced about fifty per cent of the world's steel. At that time, England had a large number of first-rate engineers, who were called 'scientists'. Japan must strive consciously to do the same thing.

'So you feel that Japan doesn't have an effective system to produce first-rate engineers?'

It may sound strange to a westerner, but this is the case. In fact, the number of high-school students studying physics and chemistry has fallen in the last ten years, from forty per cent to only twenty per cent. Our technological base is shrinking fast.

'Do you have an answer to the problem?'

Well, I'll begin with the basics. First of all, we must pay our engineers more. Our young people are in danger of losing the capacity for self-sacrifice and loyalty. This is not surprising. In Japan, society tends to swallow up inventions without even acknowledging the inventor. Just imagine, some companies have the extreme idea that engineers might just as well retire at thirty-five! Then, we should reform the education system. Particularly at middle and senior high-school level, we need to communicate the fact that science and technology are fascinating subjects. In Japanese schools today, technology is treated as a part of domestic science! And also, the university postgraduate system needs a complete overhaul. Finally, there is the problem of the social values I mentioned earlier. Technology is the basis of civilization, and is culturally important in its own right. We must make sure that society as a whole accepts this view.

'You seem then to be very concerned about the future of technology.'

If things go on as they are doing, all the progress we've made in the last hundred years will come to nothing.

'Today, we must see Japan's problems from an international perspective. What are the differences between Japan and the rest of the world?'

The American research establishment focuses sharply on frontiers. In Japan, our leading national universities are largely education-oriented.

'So you think Japan is at a disadvantage in advanced fields?'

Well, to start with, it's hard for us to get sufficient funding. We can't pay postgraduates enough to keep them interested! And in the national universities especially, there isn't much spirit of free competition; people are slow to adapt to change. Our approach is more education-oriented. In America, on the other hand, it is becoming clear that unless universities place more emphasis on educating engineers in the working areas the way we do, their industry will continue to go downhill. One of the reasons that Japanese industry became successful is because of its success in educating engineers on the job. This, I believe, is partly due to the success of our university education system, which produces people whom industry can easily train further. On top of this, it has produced world-class research. But unless more people go into research in frontier fields, I fear the country might grind to a halt. This problem calls for major surgery.

'But surely if everyone went into the frontier fields, Japan would be just like America?'

We need to strike a balance between the frontier areas and the working areas. However the habit of diligence is widespread in our culture and I think this comes from the influence of Confucianism. This philosophy was born in densely populated areas and thus emphasizes the importance of cooperation in achieving a common goal. Western culture on the other hand is a culture of thinly populated areas, and is better suited to individual endeavour than cooperative effort. When you work on your own, you are free to think things through in isolation, and this makes it easier to simplify, which is what is needed for scientific work. Science is a collection of simplifications, you see. Thus Western culture seems to be suited to science. East Asian culture, on the other hand, seems to be suited to making things collectively, although there is a danger here of oversimplification.

'Talking about the East Asian culture, do you think we get along well with our neighbouring countries?'

Well, yes I do. I suppose similar attitudes are common in countries such as China, Korea, Taiwan, Singapore, and Hong Kong. In 1988, the University of Seoul in Korea opened a new LSI semiconductor research laboratory. I went there the day after the opening ceremony. They really have wonderful facilities there. We have nothing on such a scale in Japan, not at least for LSI research. The whole lot was paid for by Korean industry, and almost everything except the measuring instruments was made in Korea.

'Korea is such a vigorous country, isn't it. One steel manufacturer has even built its own university!'

Yes, I understand that all the open public areas on the campus are paved with granite.

'Are you attracted by ideas like Confucianism or Buddhism?'

When I was a postgraduate I wondered whether I was really cut out to be a researcher, whether I had the ability. At that time I read a lot of books on Buddhism and Christianity. I even read 'A historian's approach to religion' by Arnold Toynbee. What I learned from the books on Buddhism, to put it very simply, was that you must try hard in life, even if you don't succeed. All our troubles come from the fact that we want to succeed, to do better than others. But you can be happy if you simply devote yourself to your work and this work contributes to the good of society, that is, if your work-ethic and your view of life coincide. 'Do all that is humanly possible and wait for a message from above' is a Confucian saying, but it has a Buddhist ring to it too. I believe that the main lesson of Mahayana Buddhism is that if you devote yourself to working for others, all your troubles will go away.

'Did Buddhist thinking, for example its cosmology and metaphysics, influence your research? Some advocates of the "new science"—Capra, for instance, in his *Tao of Physics*—in the West, say that "Western" science could learn from Asian wisdom.'

No, I've never been influenced in my scientific work by Buddhist or any other Eastern ideas. The way I approach science is quite independent of my personal beliefs,... or rather feelings. I don't think there can be 'Buddhist science'!

'What are your views on those relatively new Japanese forms of Buddhism such as the Jodo-shinshu, the Lotus sect, or Zen Buddhism, which were established after the thirteenth century?'

I dislike them. I only like history up to the stone age! I am interested in the origins of man, the beginnings of agriculture, the beginnings of civilization, of language, things like that. For instance, in Chinese history, I begin to lose all interest from about the Tang dynasty. The furthest I like to go is the Han dynasty.

'Why is that?'

Maybe I'm old-fashioned. There is also the obvious reason that I've never had enough time to read all the way up to modern civilization!

'Perhaps you find too much triviality in recent history?'

Yes, the driving force in early history seems essentially much less materialistic. Maybe it is a little romantic to assume this, because naturally everything trivial in those times has been washed away after tens of centuries, but all the same I feel people were more spiritual before the time of Christ. This prejudice may be why I don't enjoy reading about recent history very much.

SAKAKI

THREE

THE LAST FRONTIER OF ELECTRONICS

HIOROYUKI SAKAKI

SEMICONDUCTORS, on which the whole technological edifice of the late twentieth century might be said to rest, are essentially just one class of physical materials with strange electric characteristics. Their transformation into transistors has turned out to be one of the greatest inventions in the history of technology. It should be noted that this crucial breakthrough was made possible only with the parallel development of quantum physics, another great scientific breakthrough. The technological development of semiconductor devices can be seen as the application of quantum physics to electronics. Many researchers and technologists, however, feel that recent semiconductor devices have still not exploited quantum effects to their limit.

Professor Sakaki, at the University of Tokyo, is one of the most important semiconductor physicists in the world. He has pioneered the application of quantum phenomena to semiconductor devices (quantum devices). He has concentrated on creating a new dimension in the fundamental structure of semiconductors. Contrary to the conventional idea of striving for 'higher dimensions', he has tried to restrict the freedom of electrons to lower dimensions, beginning with three dimensions, and then down to two, one, and ultimately zero.

In this interview, he focuses on the development of his most important idea, the quantum box, which he believes is the last frontier of electronics, because in his proposed structure, not groups of electrons but one single electron will become the object of human manipulation.

———

'Were you interested in electronics and other hobbies when you were a child?'

Not at all. I never even constructed a crystal radio by myself; I was more interested in playing outside.

'That is surprising. Could you give us some examples?'

I was brought up just after the war. I sumo-wrestled and played baseball. Because I grew up near the ocean, I often went fishing. I collected sea-shells and crabs; I was out from morning till night.

'At what age did you become interested in science?'

Not until I got into high school. The first time I felt any interest was when I learned physics and chemistry. I was especially fascinated by physics. But history and English were more to my taste. My ambition was to become an ambassador. And then I had an experience which greatly changed my attitude to life.

'What was that?'

I was selected to study in the USA through a programme called the American Field Service. About two thousand foreign students were invited to American high schools. I went to a school in St Pauls, Minnesota.

'Were you interested in foreign countries at that time?'

Yes, I was. I may have been influenced by my father who went alone to Germany in 1952 and stayed for two years. I was at an elementary school at that time. I enjoyed the postcards he sent to me a good deal. My interest in English increased when I started learning it at junior high school and as a result I daydreamed about going abroad. At that time, the only chance a boy from a middle class family could get to go abroad was through AFS.

'How was your year in Minnesota?'

The most memorable of my life. The city was extremely beautiful and everything I learnt at school was fresh. I was very impressed by the way the school was run... I mean the way it emphasized the individuality of its students. For example, there was a class for students who were not very competent at maths, but also there was one for very advanced students. That was in 1962. America was a rich country and enthusiastic about science and technology as it tried to catch up with the USSR after the 'Sputnik shock'. There were many innovative experiments in maths and physics education.

'America was shocked by the Russian achievement and was trying to restructure its science edcuation... for example, the PSSC textbooks of physics.'

In mathematics, there was a programme called SMMG which was the fruit of the efforts of Yale University. I took that course and was most impressed by the attitude of students in my class. In America, people are not ashamed of admitting they don't understand something. In asking they look for an explanation which can be fitted to their own way of thinking. There's no shame in it. Even when asking a question means exposing their ignorance, it may become a way of checking the correctness of the other person's argument. So, ignorance may even be something to be proud of! You could say that if someone doesn't understand something, it may be seen as a sign of originality and therefore a strength not a weakness. So the more original the person is, the more he insists that he doesn't understand something.

'You seem to be very happy about that.'

Yes, I really enjoyed that atmosphere in American high schools. Over there the system had to serve the students, whereas in Japan people have to adjust to the system. It was the antithesis of Japan and so fresh for me.

'The early sixties was the time before the social conflicts in the USA presented serious problems. It was the golden age of America, wasn't it?'

Kennedy was President; the atmosphere was clear and constructive. The civil rights movement was in its early stages and people were proud of promoting social justice. Kennedy launched the 'Peace Corps' to aid developing countries. There were new satellites in the sky every year. It seemed to be the beginning of a new era.

'Did this year in the States change your ambitions?'

It wasn't clear to me at the time, but looking back now, two things were really impressed on me. First, as I made many friends around the world, I was forced to reflect on my abilities in an international context. I recognized that Japanese students including me were handicapped by language and that we were not very good at strong leadership. My dream of becoming an ambassador was more or less ruined.

'You feel your interest in science was encouraged in the States?'

Right. On the other hand, I became quite confident about the ability of the Japanese in maths and physics. The second thing which was impressed on me was that by becoming professionally involved in technology I could contribute something useful to the world. This was after I became aware of the activities of that Peace Corps. To be an ambassador is of course one profession, but I thought that it might be more useful if I made a contribution through something more concrete, like civil engineering. I thought this might benefit the peoples of South-East Asia a great deal.

'So your inclination towards engineering began at that time.'

Yes, I think so. Had I not been to the States, I might have entered Law School, or the Department of International Relations.

'I think it was during the time of massive student confrontation that you entered the graduate school in the Department of Engineering at Tokyo University?'

I was not an 'activist' as such. But at that time there was no clear borderline between who was an activist and who was not. It is true that everyone discussed the aims and ideals of the university, and of science in general. For example, the most 'political' members of the department, those on the urban planning course, pointed out that there was an unfavourable relationship between the universities and industry, giving the example of an academic authority endorsing industries which create pollution. The most famous example was a civil engineering specialist arguing that there was no problem discharging the waste of paper manufacturers in the Suruga Bay at the foot of Mt Fuji.

'How did he justify that?'

According to him, the waste could be naturally recycled. In fact, what happened was an accumulation of a vast amount of sludge, which destroyed both the landscape and people's health. These arguments made me think seriously about the relationship between scientific research and society in general.

'Did this experience change your attitude towards science?'

No. On the contrary, my own subject was not affected by those factors. Looking back now, my scientific 'romanticism' was not so much undermined as enhanced by the experience. I was convinced that the most important thing is not the way reputations and status generate authority, but the purely scientific attitude to the exploration of the truth. We must not authorize anything which is scientifically unsound.

'Had you already begun studying transistors at that time?'

Let me explain the problems we were having at that time. Dr Schrieffer, the Nobel laureate for the theory of superconductivity, said that taking into account the very thin conduction layer in semiconductors, that quantum effects in electrons, though possible in theory, would be negligible in practice at room temperature. I thought I had verified his prediction when I interpreted the results of an experiment on electron diffraction at room temperature while neglecting quantum effects. When I gave a talk on this result, Dr Takeishi, who is now the President of the VLSI Institute, stood up and argued that my study was pointless unless I first established that quantum effects are really negligible at room temperature.

'Were you offended by this?'

His comments rather unnerved me, but it was actually a quite sensible argument. Thus, I quickly experimented to see if there really were quantum effects when the MOS transistor was working at room temperature. It was quite difficult. When you are working at extremely low temperatures, there are many different techniques to verify quantum effects. But these techniques are unworkable at high temperatures. So I then tried to devise a new technique.

'What was the result?'

I succeeded in establishing that there were quantum effects at high temperatures by using that technique.

'So you were the first to detect quantum effects in MOS (metal oxide semiconductor) transistors at room temperature.'

It was around 1970 when I published the results. Quantum effects at extremely low temperatures had been observed by a researcher at IBM in around 1966. My discovery was significant because I disproved Schrieffer's prediction that the quantum effect would perish at high temperatures.

'It was still there at high temperatures.'

Yes. Usually, quantum effects in condensed matter disappear when the temperature rises. So it was a rough rule that the behaviour of condensed matter was described by classical physics. People simply assumed that it would be the same for MOS. But it was not.

'So, the next thing to do, as an engineering scientist, was to see if the effect could be applied to something useful?'

There were two possibilities. First, you could use it in optimizing the design of transistors. An actual consequence was MOSFET (field effect transistor) technology in silicon. Another possibility was to exploit the effect itself for useful new electronic devices. My thinking as a postgraduate didn't go that far though.

'But you were already determined to go into "quantum electronics", right?'

It was at that time that Dr Esaki's[1] paper on superlattices appeared. I was very impressed by it. To study silicon MOS one had to examine conduction of

---

[1] Japanese physicist, a Nobel laureate for the discovery of the tunnel effect.

electrons confined in a very thin film. Superlattices are multi-layered structures of ultra-thin films. His idea was to make use of vertical electron jumps between the layers which would be allowed by the tunnel effect. It was a great idea. So we talked enthusiastically when he visited my lab. He was also impressed by my research on quantum effects in MOS.

'So how did you further your research after this discussion with Dr Esaki?'

Encouraged by Dr Esaki, I went on to think of ways of engineering the electron wave by confining it in an ultra-thin film structure like that in the conduction layer of an MOS transistor. I wondered what would happen if I put a linear barrier in the film which could reflect the electron wave. I saw that when we created these barriers periodically, we could restrict the motion of electrons to only the direction of the barriers. I even thought that we might prohibit motion of the electrons in all directions and create a 'quantum box'. Most important of all, I noticed that electrons in quantum line structures will conduct in a non-linear manner as a result of electron wave interference which arises because of the tunnel effect. We could control this non-linear conduction effect by manoeuvring the gate electrode.

'Was anyone else thinking about doing this at the time?'

It was a fairly advanced idea. It was only at the time that quantum effects were being observed in superlattices. No one had thought of engraving such ultrafine structures (at the scale of 100 angstroms) on the ultra-thin film of semiconductors and, by doing that, controlling electrons. I published a paper on non-linear transport devices that could be controlled by a gate electrode. It was the first ever proposal for devices using quantum wire or quantum box effect, and it has been referred to quite often in subsequent literature.

'Am I right in saying you carried on your research with Dr Esaki at IBM?'

Yes. While working with Dr Esaki's group, two directions emerged in my research. One was the study of the characteristics of superlattices. This direction resulted in the invention of the high electron mobility transistor by Fujitsu. It is widely used in broadcast and communication satellites and is indispensible in semiconductor lasers. I would add that the infra-red detector now used works on these principles also.

'What was the other direction you took?'

As I have said, to control electrons by the devices generally called 'planar quantum structures'. This is the general term, based on my proposal, for thin films on which artificial structures capable of controlling electrons have been engraved. Thus, starting with the ultra-thin structure of semiconductors like MOS, on one hand I layered them into superlattices, and on the other inscribed an ultra-fine linear and cubic structures on them. These were the ways my research developed.

'What was the relationship between the two?'

Until some time ago, I put about eighty per cent of my energy into the first direction. It was easier to work on, and has direct applications everywhere. The second direction was extremely difficult to realize and the research was very theoretical. However, because the Japanese government backed my research on 'planar quantum structures', this area has subsequently developed very rapidly.

'You proposed this in 1975?'

Yes, I was the first to propose both the concept and its design principles. I had to wait another six years or so before people followed my ideas.

'What is the main problem in quantum stereostructures?'

There are two aspects to it. First, you have to rack your brains how actually to build a structure on a scale of 100 angstroms. Second, you have to find out what can be done using such a structure.

'Could you explain the first point?'

Scientists were trying to find a way to make ultrafine lines—or wires and boxes—by using spontaneous or natural processes. In other words, it's extremely laborious if you first make a film and then try to divide it into a fine pattern. It's more desirable to generate the pattern spontaneously.

'How is that done?'

There are several approaches. The first is the one I proposed in 1980. Another was proposed in 1984 by Dr Petrov, my friend at Bell Laboratories. I won't go into detail now. There are ways of drawing ultra-fine lines; these are usually called quantum wires. The problem now is how to create a quantum box.

'How small is that?'

Imagine cubes 100 angstroms high! We have to make them uniform, arrange them regularly in line, and fill the space between the cubes with another kind of semiconductor of very high quality. They are like artificial atoms, or molecules. If you make a bigger box, say 1000 angstroms, the electrons would behave like a lot of children in a gym. The box is too large; they can play around! You have to have a very small compartment so that the child sits quietly and properly. The ultimate quantum box is like reserving one compartment for one electron.

'In a thin film, electrons can move in two dimensions, right?'

Exactly. Classical electrons can move in three-dimensional space. In a thin film, just one degree of freedom is taken away and so they are two-thirds free and classical. In a quantum wire, they are only one-third free. But in the quantum box, there are no degrees of freedom for an electron! So we can expect it to behave as though it's confined in an atom. It is in this sense that the quantum box is an artificial atom.

'So superlattices and film structures were just preliminary stages on the way to the quantum box.'

That's right.

'Can I ask what will be possible if we build such a structure?'

Let's start with the quantum wire. Imagine that it is a semiconductor with a diameter of about 100 angstroms. If you put an electron in it, it's like putting a ball in a tube; the size is similar to the quantum wavelength of an electron. The electron can go forwards or backwards, but it can't go up and down. It can't accelerate except back and forth. In other words, it's a very interesting form of conduction because there is no diffraction. It's like an electronic bobsleigh.

'Has it already been realized technically?'

Not in actuality. But I predicted the possibility and so proposed its application to ultra-high speed transistors. We will have to wait for some time for the development of ultra-precision engineering which can actually make such a bobsleigh tube. Its property, rushing forward without any diffraction, is ideal for observing and utilizing interference in electron waves.

'What is the size of the electron cloud—about 100 angstroms?'

Strangely enough, electrons are quite flexible in size. High-energy electrons can shrink without limit! The faster they run, the shorter the wavelength becomes. The velocity of electrons running in semiconductors is about 100 kilometres per second on average. The wavelength of electrons at this speed is about 100 angstroms.

'This is much slower than light, isn't it?'

Very much slower.

'Why is that?'

Because electrons in semiconductors usually have some thermal energy when they are running. If you operate them *in vacuo*, you can accelerate them as you like, so you can reduce the wavelength even to one angstrom!

'What will happen in the quantum box? You might say it is like a gaol.'

Using my analogy of an electron being like a ball, it is indeed shut up in a tiny box of its own size. It can't move in any direction.

'So it is completely in gaol!'

If you look at it from a distance, you can't see it fluctuating thermally. It's the same as an electron which has been completely cooled to an ultra-low temperature.

'Indeed, like electrons at absolute zero, minus 273 degree Celsius.'

Yes, an electron is freed from all thermal motion there. Of course you have to have a box of only a few hundred angstroms, otherwise the electron will have room to vibrate laterally. Let's take as an example the hydrogen atom, where there is only one nucleus and one electron. The light spectrum emitted from the electron which is jumping from one orbit to another is very sharp, because the motion of the electron is precisely determined by the atom itself.

'But it's a different story in semiconductors?'

Yes, if you have aggregates of atoms as in semiconductors, the electron orbit will not be bound to each atom, because the electron or bits of adjacent atoms tend to merge. Therefore the electron orbits are easily randomized. But if you put electrons inside a box, there's no randomness in the orbit. It's the same as in a hydrogen atom! The quantum box tends to restore purity and discreteness of the electron orbits, just like the emission spectrum from an isolated hydrogen atom, even when the box is inside aggregated or structured atoms. What we are aiming at is to maintain the advantage that is gained by having a collection of atoms while restoring to them the useful characteristics which one atom has and which are lost when atoms are collected and bound together in ordinary materials.

'So, can it be said that you are trying to create a type of material where the useful characteristics of both a single atom and collected atoms are merged?'

Yes, we are. We are sort of trying to create a material state which has never before existed in nature.

'What sort of applications are there for the quantum box?'

There may be various applications, for example, lasers, photodetectors, and conduction devices using tunnel effects.

'Could you talk about these in more detail?'

In the case of the quantum wire, there will be quantum-wire transistors. All the motions except one along the wire are prohibited. So, it resembles a 'frozen' transistor except that it involves one direction only. It has extremely high speed and has low noise. This is an extremely simple application of the quantum wire. You just line them up on a plane, and then control the number of electrons by a gate-electrode. For a rather more sophisticated application, if you put many quantum wires and boxes in the emission layers of lasers, you could expect the emission of a very sharp spectrum even when the electron current coming in is very weak. This is in contrast to the present lasers, where you can get an emission of a sharp spectrum only by pumping in a lot of electric current. Using a quantum wire or box, because there is no broadening of the spectrum, you can get both efficiency and a high-fidelity light source. We could use them for a very high-speed quantum-wave inter-ference transistor (QWINT), for non-linear optic materials, for optoelec-tronic materials, and for detector materials for light at a range of wavelengths which have up till now been very difficult to detect. In other words, we will be able to obtain a vast family of quantum devices. The applications will be unlimited.

'Now can we get back to the down-to-earth problem of how difficult it is to make these devices?'

It is a very challenging field. Roughly speaking, there are two approaches. One is a sort of frontal attack. This is to exploit the technique of lithography to its limit. As you know, this is the conventional way of manufacturing an integrated circuit; it is much like etching. In this way we can make a box of about 200 angstroms. But you get defects on the surface of crystals where you cut through them. It is like a box with a lot of cracks.

'So you're not in favour of this frontal attack?'

No, I'm not. I favour another approach, which utilizes crystalline growth technology. One way to make quantum wires and boxes, which was pro-posed by me, is to expose the edge of an ultrathin layered structure and form the wire on its edge. It is called 'edge wire structure'.

'Utilizing the superlattice?'

Yes, you cut across a multilayered structure such as a superlattice and expose the edge. After that you deposit a positively charged semiconductor film on to it to attract electrons on to the surface of a specific layer. The movement of those electrons is like the behaviour of electrons confined in a quantum tube or wire. We call this method 'quantum edge wire'. It has been tried in NTT and ATT with some success. The other technique utilizes ridge structures on the patterned surface of semiconductors.

'Is this the one you have already mentioned?'

It was proposed by my friend at Bell Laboratories in 1984 and now holds some promise.

'The MITI (Ministry of International Trade and Industry) project finished in 1993 was an attempt to explore the second direction.'

Yes, to do that was one of the main objects of the project. Thanks to this project—it is called the 'quantum wave project'—I think we are now at the point where we have a clear view of the summit.

'You have said that the quantum box is "the last frontier of electronics". Do you mean that we will have nothing to do in electronics after it has been realized?'

Well, first you have electrons moving freely in three dimensions, then you limit them to two dimensions, then one dimension, and finally to zero dimension. Theoretically, you can go no further in this universe! In this sense, there's nowhere for electronics to go beyond electrons in zero-dimension. This is not to say that there's nothing more to learn about electrons. All we have done is to establish their final stage. We can't predict how they will behave on this stage. We may encounter events which are totally unthinkable to us now.

'Is there anything like the quantum box in nature? Is it completely artificial?'

That is an interesting question. Though it may sound paradoxical, each atom is a sort of quantum box in which electrons are trapped, as I explained before when I took the hydrogen atom as an example.

'So a quantum box is like an atom which has been insulated by a vacuum?'

Yes. You can't manipulate these electrons trapped by an isolated atom.

'On the other hand, if atoms are organized like they are in a metal, the electrons are not trapped around any one particular atoms. That means you can't manipulate them quantum mechanically, right?'

Yes. The quantum box created in a semiconductor gives you the chance to 'connect a wire' to a single atom, and make the electron around it an object of manipulation.

'People are inclined nowadays to call this approach "atom craft".'

It is a promising new field in engineering, and may even be the last frontier of technology. This is not an easy discipline though. We are at an early stage where we must try to systematize several technologies. The situation is rather like that encountered by architects in medieval Europe who constructed great churches over the centuries as the disciplines of mechanical and civil engineering were systematized. This systematization has just started. My contribution was just to put forth a very rough blueprint.

'What would be the main impact of this technology?'

I can't predict precisely what it would be, because we have to consider many secondary applications or inventions relevant to this 'atom craft'. I will, therefore, describe the fundamental principle of its impact. To store information, it is ultimately to mark physical space with 'something'. As you see, writing with a pencil marks the physical surface of paper with innumerable carbon atoms. But there's something much lighter in the universe which makes marks.

'You mean electrons?'

Yes. Of course. What is lighter than light? It has no mass! But you can't pin it down. It always escapes! So light is much more useful for communication. In fact, the main means of communication in the future will certainly use light. At the moment conventional computers use masses of electrons or on and off electric currents to store information. But think what would happen if we could manipulate not only masses of electrons, but a single one shut up in a quantum box. It would be the ultimate way to mark the universe. You could store information through the presence or absence of a single electron. I can't imagine what would happen if it was possible to write using single electrons.

'The ultimate in information technology?'

You have heard it said that the human brain is not an electronic computer, but a quantum computer which utilizes the properties of quanta. I suggest that, by using atom craft technology, we could bring information technology close to the workings of the human brain! You asked me if the quantum box is the ultimate in electronics and I answered sceptically. But information technology using the quantum box may actually become the ultimate way to store and manipulate information.

NAGAO

FOUR

THE LIMITS OF COMPUTATION

MAKOTO NAGAO

THE IMPACT of the electronic computer in recent history has not yet been fully evaluated. It was conceived by Alan Turing, the brilliant British logician, and was subsequently developed in the USA from the ideas of von Neumann to become the major tool of the 'Information Age'. There have been much interest, speculation, and also confusion about what computers can or should do in future. The most controversial question relating to artificial intelligence [AI] was whether computers could replace that most human of activities: thought. Leaving aside such metaphysical speculation, there have been two major challenges for the present (von Neumann type) computers: pattern recognition and natural language processing. It is essential for computers, if they are to claim any sort of parity with the human brain, that they should address these two problems.

Professor Nagao is one of the leading researchers in these fields. He graduated from Kyoto University and has subsequently pioneered the challenging fields of machine translation and visual pattern recognition.

In this interview, he recalls the initial development of computer science in Japan, and then goes on to describe and discuss the development of natural language processing by computers, specifically machine translation, from his uniquely philosophical viewpoint.

'Professor Nagao, I understand that when you went into electronic engineering, the field was in its infancy?'

Yes. In fact, the word 'electronics' hadn't been introduced into Japanese yet. The year I took the entrance exams for Kyoto University, they had just set up the department of electronic engineering, which was the first in Japan. I didn't have much idea what electronic engineering was, but because it was new and I thought it might be interesting, I applied. In fact, I honestly didn't know whether I wanted to enter the science faculty or the engineering faculty, or even whether I should take up philosophy. I was very interested in the humanities. My father was a Shinto priest and I had learned a lot about Shintoism from him. I was quite interested in spiritual issues at that time, so it was difficult to decide.

'So you might have followed in your father's footsteps?'

If Japan hadn't lost the war, I think I might well have done, because there was a strong tradition of succession at that time. But at the end of the war my father said, 'Times have changed, do what you like'. So I did what I liked, and here I am. But I was very interested in spiritual things at the time, not just in Shintoism but other, more transcendental, concerns.

'And are you still interested?'

Yes. The *Kojiki*[1] is fascinating, and the *Manyoshu*[2] is truly great literature.

'It strikes me that language can be approached from the point of view of computers, and at the same time as a cultural, semantic, or philosophical problem. Given this overlap, the fact that you were interested both in the humanities and the sciences from high school age was an excellent preparation. But in the end you settled for the sciences?'

The physical sciences and maths were my strong subjects at high school. But to be a really good mathematician, you have to be able to look at things in a simplified way. You have to have a dry mind to be able to do this. In contrast, I see my mind as rather 'wet', rather indecisive, that is I tended to see an object not in an analytical way but in its totality. Taking pleasure in discovering the simplified principles underlying something may be interesting, but I feel that one should be able to explain the parts that conform to these principles. It is much more important to explore the elements that can't be explained by basic principles, those illogical elements which I think are the real reason for existence.

'I believe you belong to the first generation of pure computer scientists in Japan. When you took up the subject at the end of the 1950s, what were you first interested in?'

I wanted to see what computers could be made to do. I was interested in finding out how far computers could reproduce the functions of the human mind, so I decided to tackle the particularly human problems of pattern recognition and natural language processing.

'Both those themes are concerned with the way in which human beings recognize things?'

---

[1] The mythical history of Japan written in the seventh century.
[2] An anthology of old Japanese poetry, compiled in the eighth century.

They are connected by the basic question of whether it is possible to devise a computer model for the human recognition process and for translation mechanisms. These fields were then very young and the engineering models that existed at the time were limited by engineering theory and practical considerations, so they came nowhere near to replicating human brain processes. From this point of view, the models we came to devise were quite different. By 1963 we had already worked out a framework, though a very simple one, for a model that we thought might be suitable for the machine translation of language.

You have to remember that computers in those days were like toys compared with recent models, and their dictionary was considered big if it contained two or three hundred words! The grammatical theory we used was quite primitive too. Anyway, we have struggled on, and recently I have begun to feel that we may have reached the point where our system might possibly be realized on a practical basis. In those early days, by the way, there was a lot of discussion about the ambiguity of language. Some researchers proudly claimed to have identified ten or twenty different possible interpretations of a single sentence! This was nonsense to me, though. Why couldn't they just establish useful parameters for eliminating nonsensical interpretations? An important clue was 'meaning'. But nobody really looked into this. I suppose it must have seemed like an impenetrable jungle.

'How did you approach it initially?'

I like to think of a sentence as a tree. Grammar is the set of rules which specifies the structure—the tree—of a sentence. Meaning is the system which supports the co-existence of words—you might say the leaves—in a sentence. To use another image, grammar is the warp and meaning is the weft in a piece of cloth; both are necessary and complement each other. The warp unifies the sentence. I felt that the introduction of these constraints would help us extract a meaningful interpretation from the many structural ambiguities.

'But you had to realize that view with actual programs?'

To test my theory, I wrote a program for sentence generation, inserting consistency of meaning as a condition. Grammar on its own would produce a lot of meaningless phrases, but by introducing this application of meaning it would be possible to create sentences which preserved compatibility between their words, and thus create only meaningful sentences, demonstrating the validity of using these constraints.

This work attracted a great deal of attention at the first Conference on Computational Linguistics in New York in 1965, and I was very warmly invited to continue my research in America.

'Did you go?'

No, I didn't.

'Why not?'

In 1962 I was invited to an international conference on Algol 60, a computer language, which was held in Rome. Afterwards I toured Europe and came away very impressed with European culture and European thought. The USA had become the centre of research in my field but I was deeply impressed by the fact that the Europeans had their own way of thinking and their own methodology and that they pursued original research, even if the results seemed unspectacular.

So I thought that if I was going to do research abroad, it would be in Europe. And then I thought since the facilities for research in the USA were so much better than those in Japan or Europe, I might go to America after all. Finally I decided that good creative work could be done anywhere; Japan was as good as anywhere else. I felt that to be honest in my motives, I should stay in Japan and work for my country, and even if the conditions were bad I would just have to persevere. So in the end I refused the invitation from America out of pride. I really don't know if what I did was a good thing or not.

'What happened to your research on machine translation after that?'

Well, from 1965, I continued to work on small-scale machine translation systems. We constructed many English-to-Japanese and Japanese-to-English systems and discarded them over and over again. This went on for about ten years with some success and much disappointment. Machine translation was developing poorly. When we had German or French research students in our laboratory, we worked with their languages.

'It seems far more difficult to work with such different languages as Japanese and English.'

Yes, the principle common to all these systems for translation, as I said earlier, was a method based on grammar which made as much use as possible of information about meaning. A machine translation system based on grammar with a context-free phrase structure was not powerful enough to obtain a sufficient grasp of the relevant linguistic phenomena. So simple a framework is inadequate especially when you're translating between two languages whose structures are as different as those of Japanese and English. We devised a model which used context-sensitive grammar, which has more expressive power than context-free grammar, and we added to this the checking function of applying consistency between the meaning of words. This would correspond to Fillmore's case grammar concept which takes into account the semantic relationship between subject and verb.

I was already looking at this framework when I decided to work on machine translation in 1963, and even today, from an engineering perspective, it is probably the most appropriate framework for a machine translation system.

'Would you say it is very close to human thinking?'

It's not a particularly 'human' system but, from an engineering point of view, it is robust and relatively easy to improve stage by stage. Similar systems are being developed all over the world now.

'Are there difficulties building a working machine translation system in a university?'

There are indeed. You can only produce the basic technology of machine translation and only small-scale models in university laboratories. These are suitable for research theses in university laboratories. It is extremely difficult to build anything on a serious scale in a university, especially in Japan. But when you're dealing with something as complicated as language, it's no good simply picking out those bits that fit the theory beautifully and crowing over them. You only get an idea of the difficulties of the problem if you handle the large variety of expressions which are used in real life. This is true especially because that proportion of language which is irregular and unique is far larger than what can be explained by grammatical rules.

'So you always had to deal with language as a "whole", and that made you evolve more and more complex systems. That's the reason you were involved in the famous government machine translation project.'

Our toy-like systems were quite inadequate and we were constantly aware that we would one day have to build a large-scale system. So we tried to build a usable English-to-Japanese translation system to translate scientific and technical research papers.

This system was successful and the Science and Technology Agency (STA) of Japan was interested in it. The Agency asked me to start a research and development project for the large scale machine translation of abstracts of science and technology papers under its auspices. I accepted the invitation and for four years from 1982 we worked on building Japanese-to-English and English-to-Japanese translation systems for dealing with these abstracts.

'Didn't it seem too big a project for an academic researcher?'

It needed a certain amount of courage to take it on. There was nobody else in Japanese academia with the experience or know-how to tackle it. But we persevered and, within the four years allocated to the project, we managed to complete two systems capable of translating the abstracts of scientific and technical research papers, one from Japanese to English, the other from English to Japanese, and the results were for the most part well received. Afterwards, the system was improved and polished up and today it's in daily use at the Japan Information Center of Science and Technology. So this project played an important role in the development of machine translation in Japan and helped to make Japanese machine translation technology the best in the world.

'So if a manufacturing company wanted foreign research papers translated, your system could be of use?'

Yes. The system is basically designed to cope with any kind of text, but when it actually comes to using it for a specific purpose, it needs to be customized by augmenting its dictionary and grammar. However, once the system has been customized for a special field, it is rather difficult to use it for a different purpose, because there is always a gap between one field and another. So then you have the problem of bridging the gap. It is the problem of specialization. The more a system is specialized, the more precise it becomes, but the less flexible. So how do you make such a system more flexible? That's the big question for the future.

'I think this is a general problem in modern research.'

Yes, the problem is not specific to machine translation alone. Consider this: in Japan, we have *furoshiki*.[3] In the West, they have briefcases for books, suitcases for clothes, and so on. There is a different container for every purpose. In itself, this is a very convenient system, but if you try to put anything slightly unusual into the bag, it won't fit. The *furoshiki*, on the other hand, doesn't excel at any single task, but it does have the advantage of being able to wrap almost anything. This is one typical example of the differences in thinking between Japan and the West, and I think that in future we will all need to develop more adaptable systems, like the *furoshiki*.

'Your work demands that you be any number of different things: philologist, grammarian, student of the English and Japanese languages.

Yes. There is, however, no such thing any more as a linguistic theory for all languages. A linguist no longer studies language as a whole; he is only interested in certain limited phenomena. In other words, the field of linguistics is already divided into an incredible number of specific sub-fields. An engineer, on the other hand, even if his framework is not so precise as these linguistic

---

[3] 'The carrying cloth', a traditional device which can wrap up anything to make it easier to carry.

theories of specific sub-fields, has to take a comprehensive view of language so that his system can guarantee a certain level of quality. So he adopts quite a different approach from the linguistic theorist. The difficulty is how to reconcile these two approaches.

'Of course; their goals are different so I suppose it's natural that their approaches should also be different. So, from an engineer's point of view, you already have some kind of independent grammatical system ready to be programmed into a computer?'

Yes, I think we have. Even if a linguistic phenomenon has to be specified in detail within linguistic terms, it does not necessarily need this distinction to be made from an engineering viewpoint. We can introduce another aspect which allows the problem to be treated by a computer.

'So machine translation is already well on the way to becoming a reality?'

There are many commercial systems being used now in Japan for the translation of technical literature. But I want to make it quite clear that the systems on the market today are still unsatisfactory. By the beginning of the next century, there will be a machine translation system with a quite different concept and framework. I should hasten to add, though, that this will be for translation of written texts, not of speech. Even this translation machine, although it will be of high quality, will have difficulty in rendering the very human nuances of linguistic context. The thing about human beings is that, provided that they have an idea, words just seem to follow on from it. Machines, on the other hand, chase after words; they can only deal with words as words. And that is no use at all.

'Would it be possible to build a robot and give it some kind of "experience"?'

Human beings 'experience' things twenty-four hours a day, even before they are born. And most of this experience is new because the world is unpredictable; so they are always learning. How can a machine ever aspire to this kind of experience?

'How about your research on pattern recognition? How did that develop?'

Well, it started mostly in America and developed as a practical application of a theory known as statistical decision theory. This approach, in the area of character recognition, gave some excellent results. That was in the second half of the sixties.

'I thought you were always critical of this approach?'

As I have said, I liked maths when I was at senior high school, but after I studied pattern recognition some time around 1960 I began to have doubts about whether human beings could really recognize patterns using a mathematical process, and so I lost interest in a methodology which used maths as a means of interpreting everything. There are many aspects of pattern recognition that cannot be described by statistical decision theory, and it was these which claimed my attention at the time. I had embarked on quite a different route from the one being followed in America.

'Do you know why their theory failed?'

In dealing with *kanji*,[4] for instance, you have to take a far more structural approach to the features of each character than is possible in statistical decision theory. This is equally true when one is trying to analyse other structural aspects such as the two-dimensional relationships between features in people's faces and in ordinary landscapes. I began work on the mechanisms involved in recognizing features in this kind of image, how far it was possible to reproduce these mechanisms on a computer, and on image processing in general.

'When did the machine reading of post (ZIP) codes begin?'

I think about 1970. Toshiba used our research to build the first post code reading device, and it was a considerable success.

'So the systems you have developed are based on a structural reading of individual elements?'

That's right. My method was based on different kinds of structures which are found in numbers. It had the advantage of being very easy to commercialize.

'What is special about pattern recognition technology?'

In pattern recognition, you can't use one principle to cover all the possibilities of pattern variation. The best you can hope to achieve is about 90 per cent. To get to 95 per cent, you have to add all kinds of things to the original principle and even modify it to a certain extent. And to get to 99 or 100 per cent is impossible using just one principle. You have to apply all kinds of extra principles *ad hoc*. It's very difficult to create a system which is both mathematically neat and 99 per cent accurate, because human beings do not produce phenomena which obey natural laws. That's the reality of engineering.

'So as you approach 100 per cent accuracy the difficulty increases exponentially the closer you get?'

Yes, it does become extremely difficult. Human beings may be able to recognize simple patterns instantaneously, but for complicated patterns, we

---

[4] The Chinese characters used in the Japanese writing system.

may have to combine a lot of different methods to arrive at the most appropriate conclusion. For instance, if you are trying to recognize a sound, another noise might interfere, or if you are trying to recognize *kanji*, two may be stuck together and would have to be separated before each character could be recognized. This means that you have to use different processing techniques and decision methods to arrive at one conclusion. A computer must go through each and every one of these tasks, or it will never get anywhere.

'We were talking about the post code reading device, but in fact image processing can be used in a whole range of different fields?'

Yes, clinical image processing is one example. In medical diagnosis, it is used to sort through large numbers of X-ray photographs and pick out those which show something unusual. There are many other examples. For instance, the use of image processing is one reason why integrated circuit technology is so advanced in Japan. To make an integrated circuit, you etch a wide variety of pattern masks. Often, you have to superimpose a number of pattern masks to an accuracy of one micron or less. If they were positioned by hand, productivity would be very low. So image processing technology is used to recognize the positions of the masks and they are then automatically fine-adjusted. And image processing is used again to check the finished integrated circuit for defects. That's how Japanese firms are able to maintain such high quality in the manufacture of integrated circuits.

'*Kanji* recognition is a discipline unique to Japan, I think. You are also a pioneer in this field.'

Initially, during the sixties, I worked on *kanji* recognition for about four or five years, and then a number of electronics firms caught up with my research, so that manufacturing companies started to develop character recognition devices. Many other people in academia became interested in character recognition too, so that I thought there was no point in trying so hard any more and I gave up this field of research. Of course there still remained many interesting and difficult problems to be solved in character recognition, not only at an experimental level but for practical systems too. But I thought I should move on.

'Where did you go?'

I became interested in image processing. I analysed grey-scale images which were taken from photographs. That was in about 1965. I suppose I was one of the pioneers in this field, but it was certainly hard going.

Once a lot of people get into any particular fields, which is welcome to me, there is no real need to try to stay ahead of them. Once this happens, I get a strong urge to look for a new subject which no-one else has studied. Even now within image processing, I am studying a difficult field which hasn't been tackled by anyone else—general purpose image processing systems.

'So this is what you are focusing on. Could you explain what you do?'

I worked with photographs of people's faces and tried to develop a system which would analyse the position of the eyes, the nose, the mouth, and their mutual relationship, and then use this data to identify the shape of the face automatically. But in fact, the machine tended to make a lot of mistakes. For instance, it might identify something as an eye which might not in fact be an eye, and what it identified as a mouth might in fact be a nose! This might be because the lighting created different kinds of shadows on the faces, or because some of the subjects wore spectacles and others had beards; all of this caused interference.

'So what method did you adopt?'

The machine might find something that has, for instance, the shape of an eye. It then checks that this shape is positioned under an eyebrow and to the top right or left side of the nose, that the nose, mouth, and chin are arranged in the correct order, and that the relationships between all these features are appropriate. Only then, when the machine decides that all these features are correctly arranged, does it complete the recognition process. If it judges that the arrangement of these features is inappropriate, it then decides that one of the features has been wrongly identified and it has to go back over its tracks. You may think this is the natural recognition process used by human beings, but this is not so easily realized by machines. What is easy for a person is sometimes very difficult for a machine.

'So, essentially, you introduced the concept of feedback.'

Yes. We developed a very sophisticated programme using this kind of background knowledge about faces, designed in such a way that when a result did not tie up with the known facts, the machine decided that it had made an error of recognition. It then had to change the conditions and reanalyse the object in question. I believe that it was in 1971 when this kind of feedback process was first used in image processing anywhere in the world.

'What was the reaction at the time?'

We published our report in 1972, but people working in artificial intelligence at that time did not recognize the importance of our work, although later it came to be adopted widely without any acknowledgement. I don't complain about this. The most important thing is that something good was created and that society has benefitted from it.

'Feedback is something that is used in automatic control, isn't it? It's interesting that you have adopted it for image processing.'

I believe that the human mind goes through this kind of trial and error process when trying to recognize something complicated. We tried to make computers do something similar. We did in fact introduce another interesting method in image processing: an expert system method which used a sort of 'blackboard' process.

'Is this related to other problems than face recognition?'

When we had finished analysing photographs of people, we went on to aerial photographs and these were even more complicated. We analysed these photographs so as to be able to identify roads and rivers, where there are three houses in a row, whether this is a forest and these are fields, and so on. In photos of faces, the eyes, nose, mouth, and so on always have the same spatial relationships, but when it comes to aerial photographs, there are no reliable generalizations you can make about the relative positions of trees and fields, houses, roads, or rivers.

'In other words, you can't rely on finding a common, regular structure.'

In such cases, you have to look at local characteristics and relationships. For instance, roads are long and narrow, with many different junctions; rivers are long and narrow, their ramifications are usually Y-shaped, and they are usually pale blue. Houses are usually rectangular and tend to face on to roads; forests are deep green and hairy-looking. This is the kind of thing you look for in an aerial photo.

You try to build as many as possible of these characteristics and their interrelationships into the image processing program, so that they are used as rules by which all the possibilities can be examined. The results are written up in a special field called the 'blackboard' and are then processed, erased, and rewritten until there is a final result which satisfies all the rules without contradiction. We completed this image processing system in 1978. I believe we were the first to introduce this method into image processing.

'You said in connection with your research into face recognition that you were the first to introduce a feedback process into the field of image processing. Is this kind of system practical for use in general-purpose image processing machines?'

If you know the shape of an object in advance, it is possible to devise a procedure for processing it, incorporating some kind of feedback process. But when you're dealing with aerial photos and other images in which it is impossible to predict the general shape, you can't specify procedures. Instead, you input rules about local relationships and leave the analysis procedure to the system. The system is designed to go as far as possible by trial and error, so you try to build in as many reliable local rules as possible.

'So the machine is becoming cleverer. What comes after that?'

You could imagine a more advanced recognition machine. It may be an exaggeration to say that a general purpose system will cope with any image in a general situation, but suppose you want to recognize a coffee cup. You show it to the machine which then analyses it in its own way, and builds up its own definition of a cup. By showing the machine different objects, you build up the number of things it can recognize. Of course, with the systems we have at present, people have to help out a little to make sure that the information is organized properly, but the eventual goal is to develop a machine with a learning function that will help it do this automatically.

'Natural language processing and image processing can be defined as artificial intelligence in the wider sense, but I understand you do not think of your own research as being in artificial intelligence?'

Well, artificial intelligence became a catchphrase in America, and different people understood different things by it. If you define artificial intelligence in a very broad sense as the mechanical reproduction of human intellectual mechanisms, then my work on language processing, machine translation, and image processing could all be called research in artificial intelligence. But I prefer to say that I am a specialist in natural language processing, machine translation, and image processing. AI is too vague a concept; it is not concrete enough.

'Will machines ever have minds or consciousnesses?'

That is a big question. I don't think it will happen in the near future. The first task is to imitate human intelligence, not human emotion which is an important part of the mind.

'Can that be accomplished with the computers of today, that is with Turing machines?'

I think so, but the problem is how to transfer the very ambiguous processes of human verbal communication and image understanding into algorithms or programs, the languages that computers understand.

'Then the problem is with software?'

Perhaps more than software, the problem occurs at the previous stage, that is passing from the irregular realm of human intellectual activity through algorithms and then to software. Initially, we can develop computer systems for fields where problems are well defined and efficient problem-solving procedures can be devised. This is the science of software. Systems which have to handle more vague and ambiguous problems which can't be formalized in algorithms depend more and more on 'expert systems' whose methodology cannot guarantee a correct solution to every situation, but can produce results which are usually acceptable. This method is called heuristic. The human brain has a lot of these systems. So expert systems are gradually becoming used universally. I would say we have only just begun to understand the extent of the difficulty, the complexity, of reproducing human intelligence in computers.

'Some people say that the fascination of artificial intelligence has faded in the 1980s.'

This is certainly a period of change. Personally, I didn't set out to become a researcher in artificial intelligence. Minsky and other people seem very sure of themselves, but in America you have to shout loud and publish often, otherwise you don't survive. At any rate, people working with artificial intelligence used to boast that machines would soon be more intelligent than human beings but in fact the machines worked at a childish level. Neurocomputers may have come back into fashion but they still can't compete with the complexity of the human brain.

'Do you rather feel that people working in brain physiology are looking to neurocomputers for inspiration?'

I don't believe this, but I wonder what would happen if computers could mimic the independence and imagination of the human brain. We might find ourselves having to compete with computers. It would be just as if a new kind of brain had been invented. Would we still really understand the computer? In fact, computers appear to be catching up with human intelligence at an amazing speed, but they are only excellent in certain specific areas. For people, thought and action are a whole; our brains have unbelievable complexity. It's really a mystery. So it's possible computers will never catch up.

'You said earlier that your father was a *Shinto* priest: do you feel any interest in *Shinto* or perhaps—how shall I put it—the Japanese way of thinking?'

Yes indeed. I am still interested in *Shinto*. Of course, I don't mean the nationalistic post-Meiji[5] form of *Shinto*, but the 'philosophy' which is rooted in the Japanese way of life. In *Shinto* you are not bound by dogma unlike the totalitarianism of monotheistic religions. Instead, *Shinto* has a very 'fuzzy', flexible pluralism. This Japanese philosophy is present in all our classical

---

[5] The era from 1867–1912.

literature, for example in the *Tale of Genji*[6] and the *Heike Monogatari*,[7] but I'm starting to ramble—perhaps I'm getting old!

'Do you feel that this Japanese philosophy has influenced the development of our technology?'

Well, not directly, but if we consider artificial intelligence, our approach can be contrasted sharply with that of US researchers. They tend to set a great deal of store by abstract theory. They like to concentrate on a part of an object and then explain its mechanisms. This theory might be self-contained and impossible to refute. They take less care about the object as a whole or they intentionally avoid getting involved in seeing the complexity of the whole mechanism, which is too difficult to grasp using a simple-minded theory. They are often happy in this artificially specified world. Their theories are sophisticated but when asked how they will work out in practice, they often take the attitude that it doesn't matter. Japanese researchers think differently. In Japan, many researchers think that theory is of no use if it doesn't work in practice. It's a toy theory you might say; not so much a theory, as an assumption in a very restricted virtual world. We want to understand the whole reality or at least to grasp the object's relation to it. We can't be so simple-minded as to believe so easily in neat theories in a virtual reality. Problems are usually more complex than we think.

'Are you saying we are not on a par with the USA in fundamental attitudes to the subject?'

Perhaps we are more timid than they, or just more conscientious. At any rate, when Japanese researchers present their inventions they tend to say, in a small voice, 'Excuse me, I'm not sure how strong my theory is but it *does* work …'.

'But isn't that the true spirit of engineering?'

Yes I think it is. Engineering has an interesting characteristic. A simple theory may achieve 70 per cent accuracy or degree of correctness. Beyond that level, no good theory exists for many of the problems with which AI research is concerned. But if you spend enough time, money, and energy, you might achieve about 95 per cent. 'I don't know how it works, but it does work' is the true spirit of engineering and perhaps of the Japanese people too. We are much more interested in dealing with those difficult areas which are beyond theoretical explanation. But I am not saying that Japanese engineers don't respect theories. Of course they do. What I *am* saying is that they tend to be attracted by the problems which seem to be outside the realm of current theory.

---

[6] A famous ancient novel written in the eleventh century.
[7] A medieval story of civil war written in the thirteenth century.

'How then do you see American researchers?'

American researchers are very sharp and rational in their analysis of phenomena, in their research papers, and make strong claims for the validity of their ideas. Very often, though, this validity is restricted to ideal situations or for toy subjects. The important thing in engineering, of course, is not to publish lots of papers but to arrive at comprehensive solutions for actual situations and for a wide variety of objects.

'Your approach sounds similar to the British concept of empiricism in some ways.'

Maybe. British empiricism, although it is bound up with Christianity, is different from the philosophy of continental Europe. It is an idea with a rounded, comprehensive approach. But this approach is actually more specific to Eastern thought, I suppose. Take for example the practitioners of Chinese medicine. They try to see human beings as a whole. It's not an analytical medicine which focuses on the disease and not the patient.

'Science and technology as we know them today originated in the West and spread over several hundred years through Asia to Japan and now hold sway over the whole world. How do you see them developing in the future?'

We have come to the point where science and technology must struggle to produce something new or something which is truly valuable for the future of human beings and their environment. It may be that science and technology will only progress further with difficulty in their present form. We have to consider seriously what progress is and what it means for the future of human beings.

'So you are sceptical about current trends?'

I doubt that science and technology will continue to grow in the twenty-first century at the same speed they have grown from the beginning of the nineteenth century to the present day. I can't help wondering for how much longer the future will seem to be rosy. I suppose you could say it's questionable whether society will continue to place the same value on science and technology in the twenty-first century as it does now.

'So you are envisaging a sort of change in the value of science and technology itself?'

Well, probably. Whatever the case may be, I imagine that we will continue to regard science and technology as valuable disciplines which have made a major contribution to the well-being of mankind at least until the end of this century. However, I wonder whether this belief in science and technology will really last forever. If it doesn't continue, what will the next object of worship be? On what will humanity rest its faith? I feel this is an immensely important problem, one which I would like to pursue further.

'I wonder whether the Japanese trait of accepting things that are not fully understood may provide an unexpected solution.'

Yes, I feel the same way about it. For instance, take what has happened in Russia and Eastern Europe. There we saw how mankind's attempts to realize the ideals of communism have failed. No matter how strong the theory is, the ideas of a single human being are not so almighty. There is no solid theory that will stand up through the whole world or through the whole universe. There's nothing absolutely true in the real world. Everything is relative and the truth is bound by certain conditions.

Suppose you try to formulate a principle or rule for the natural sciences: you must consider whether it will hold up through the entire world or the entire universe from faraway galaxies to the internal space of atoms. As long as you are not able to do this, I believe you can't rest on your laurels. No matter how broad a view you take of the world, any underlying principle you may think you have discovered is bound to be only part of the picture. There will always be another world in which this principle is invalid.

In this sense, we scientists must admit the existence of something which is unknown, which cannot be determined as true or false. We must coexist with things which are contradictory in themselves.

'I am afraid your thinking might be interpreted as somewhat mystical, would you say it is actually an Eastern way of thinking?'

What I probably want to say in a nutshell is that we must respect every state of existence, or of nature itself, as it is. In this respect, the Japanese tradition of living in harmony with nature may, without any logical basis, testify to great wisdom.

'So as our wisdom increases so values change; we will then realize that science and technology are only partial solutions to our problems and we will need to review our attitude to them?'

Science and technology will no doubt continue to exist in the future, but I do feel that values are changing. In the middle ages, for instance, people had absolute faith in God. This was particularly true in the Western Christian world. In this century, however, we tend to have absolute faith in the power of scientific truth and logic. And I wonder whether that faith will continue into the twenty-first century. If not, what will we all believe in in the years to come? For me, that is one of the most interesting questions.

In the old days, one used to judge things in terms of how they measured up to God. Nowadays we ask ourselves how a phenomenon relates to the known principles of science. The real question will be whether the approach and methodology of the natural sciences continue to be the standard for social values, or for faith.

'Do you think ordinary people are satisfied with science and technology?'

I think we are becoming more sensitive to the destruction and devastation of modern society and the earth which may have been brought about by science and technology even though unintentionally. People will base their value judgements on whatever they feel. This feeling will be more important than scientific theories and truth, I suppose. What that will be, I don't know at present. That's what I'd like to find out.

'So it is an unpredictable world based on feelings and not on theories.'

Of course, the principles of science and technology are quite firmly established. I believe they will continue to develop steadily. And I believe that many things we don't understand today will be explained and that new and useful technologies will be developed. I personally intend to persevere along those lines for some time. But the other side of me does wonder. Maybe the general course of science and technology has hidden pitfalls. It's possible that we feel this unconsciously and this makes us uneasy. And I also feel that we may be on the verge of breaking into another world with quite different values.

'A universe which incorporates a very Eastern kind of "chaos"?'

Well, after all, we are living in an illogical world and we are able to hold two sets of conflicting views at the same time.

FUCHI

FIVE

THE CHALLENGE OF THE NEW GENERATION COMPUTERS

KAZUHIRO FUCHI

ELECTRONIC COMPUTERS, which have evolved into such diverse types, from supercomputers to PCs and microcomputers, share the same simple architecture as that designed by von Neumann, based on Turing's theory. Forty years after its invention, however, the limits of the von Neumann machine have become apparent to computer technologists throughout the world. It may be impossible for the current design of computers ever to accomplish the tasks set up for artificial intelligence. As a result, since about 1980, there has emerged a call for 'New Generation Computers' which could surpass the limits of the von Neumann machine. The Japanese government is one of a number of countries which have decided to invest heavily in computer research into the next century.

Professor Fuchi, the former President of the Institute for New Generation Computer Technology (ICOT), is the leader of this national project in Japan. The launching of this unique institution got much attention all over the world, not only because it was so timely but also because it symbolized the emergence of Japan as a technological superpower second only to the USA. This project ended in 1993 and Professor Fuchi has now left ICOT to become a professor at the University of Tokyo. Although the outcome of the project has had many contradictory evaluations, it has had little coverage in the foreign press. Professor Fuchi discusses in this interview the spirit of this experiment which lasted for ten years.

'The "Fifth Generation Project" that you led was terminated at the end of the 1993 academic year, and I hear that you are now leaving ICOT. How does that make you feel?'

I feel that we did our best. The young generation especially has been very enthusiastic.

'In spite of severe criticism from overseas?'

I suppose the concept of ICOT itself has not been fully understood. There has been much superficial coverage by the media.

'I think this is an important point. Could you explain in detail?'

I would like to respond to two levels of criticism. At the one level it was based upon misunderstanding. For example, it was said to be reckless to try to solve the most important problems in the AI research within ten years. No one in ICOT claimed any such thing. Also we were accused of trying to set a goal of creating a machine translation system equipped with the same capabilities as humans, and that this was destined to fail. We've never attempted such a plan. I remember there was even a very strange book which maintained that the real intention of the ICOT project was to deal with Chinese characters, Kanji.

'What do you think are the reasons for such misunderstanding?'

One very specific reason, besides a lack of research by these journalists, was that they confused the conference for the future generation computer system (FGCS '81) with ICOT. Though this conference was organized by ICOT before it was actually established as an institution to provide a forum for computer scientists around the world to propose their ideas for the progress of computers in the next century, the actual projects at ICOT were in fact much more specific, and had been well targeted from the outset.

'And what is the second level of criticism?'

Well, ICOT's slogan had 'Parallel Inference (PI)', or the 'Parallel Inference Machine (PIM)' as its hardware concept. Actually, this was all that we had promised to achieve. The second level of criticism was that we were wrong in taking this direction.

'So they actually understood what you were doing, but they were against it?'

Yes.

'So what is your response to that? In what sense have they misunderstood you?'

They perceived ICOT as a conventional Japanese research group. They thought that the basic stance of Japanese research was to apply principles established outside Japan.

'And you are saying that this was not true of ICOT?'

No it wasn't. What I really want foreigners to understand is that 'PI' was a hypothesis. It was not a dogma, though we firmly believed in it. We were not aiming at manufacturing specific products. In fact, what we did was to perform experiments to test our original hypothesis. We would have been quite happy to throw out this hypothesis if it was proved to be wrong. However, after ten years' activity this was not the case.

'So you are saying that the spirit of ICOT was purely to experiment with new ideas.'

Yes. To demonstrate that it was an experiment, I would like to point out that the idea of PI was never an accepted principle on which we could base the development and research of new applications. The idea of parallelism was in the minority when we established ICOT, and Parallel Inference is still an exotic concept to many researchers.

'It was an experiment because there was no guarantee that it would succeed.'

It was not a project aimed at the development of so-called Japanese products as journalists liked to interpret it. What we were up to was very basic experimentation and we were not at all interested in making a specific product. The criticism of leading American journalists was completely off the mark.

'Does this mean that the West's stereotype of Japan does not allow for the possibility of basic government-backed research which is not aimed at some sort of application?'

Yes, and there is another misapprehension about Japan, which is that we look for unity or if you like, a 'trinity' consisting of industry, government, and the universities in harmony, and that we imitate foreign technology and efficiently turn our knowledge of it into products. This image is completely wrong; many attempts of this sort have resulted in failure. ICOT is not this sort of organization.

'What is it actually then?'

It is true that ICOT had some fuzzy aspects which are difficult to explain to foreigners. For example, the fact that ICOT was founded on a simple and clear hypothesis is a Western rather than Japanese idea. However, the environment at ICOT was very Japanese, it is one of speechless communication among very intimate members.

'What was its financial status?'

MITI did not offer a grant to any company via ICOT, and although these companies sent us excellent people, they did not profit from it, and nor did MITI or any other company have control over us.

'So you were effectively in full control?'

Well, I would say so, although I don't know how younger people think.

'I believe the organization of ICOT was modelled on Den-So-Ken[1] of which you are an ex-researcher?'

Yes. Den-So-Ken was unique because we enjoyed there maximum freedom of research, although we received poor funding from MITI. So ICOT was effectively another Den-So-Ken for basic research.

'Projects influenced by ICOT have been tried out in the West. What has been their outcome?'

They have all failed. This is owing to the misapprehension of Japan as a Trinity of industry, government, and universities. The relationship between these groups in Japan is fuzzy, and would not succeed if it were artificially created as it is in the West. In the US, I heard a case described where researchers from different companies would not dine at the same table. In the UK, a situation arose where a number of outstanding scientists working for a poor company became unhappy with their situation, and the company refused to pay them. Imitating Japan is therefore out of the question, and those areas in which we can be imitated are unimportant.

'Your theory of the Fifth Generation Computer sounds like the sort of thing Americans are traditionally good at. They have many good original ideas and then try them out, although sometimes they leave the final touch to the Japanese. So when the ICOT experiment was announced to the world, were they against it?'

PI or PIM, though it is now at the level of practice, was only an idea when ICOT started. Since it is a matter of opinion, I have never tried to insist on my originality; I prefer to say that it was all a natural development from various lines of research. Because the stance of ICOT was not very Japanese, it may have touched on the essence of Western rationalism. Therefore, I thought the Americans would naturally accept it, although they might criticize our project in a constructive way. In fact the opposite happened. It was more like, 'who do the Japanese think they are?'.

---

[1] Electro Technical Laboratories, the MITI research institute which has a long history of electrical and electronic research in Japan.

'They didn't like to see western rationalism practiced in Japan?'

Maybe it was a sort of jealousy. Especially since the US was ahead in research and had many projects under way. So, leaving aside young researchers, those who were in the position of making proposals had to base their ideas on previous research in the US. To these people Japan was little more than another State, like California for instance. And yet it wanted to create a research project of its own. I suppose that's why there seemed to be such a great emotional reaction.

'Were there also hostile feelings because Japan undertook this futuristic Fifth Generation Project using government money?'

That was not a reason. In the US, there are projects funded by the Pentagon. So the source of the money was not a cause of hostility. It was the way basic technology was transferred to industry. They thought that the Japanese government was trying to back Japanese industries by creating a institution whose results would be enjoyed solely by its own members.

'So, what's your answer to this?'

What we are doing now is of no immediate use, nor is it a means of developing actual commercial machines. Because of this, Japanese computer manufacturers have been quite unhappy with us; we are not working only for them, but for the computer manufacturers throughout the world, wherever possible. I never thought that we were working purely for Japan's national interest. The official stance, however, was that because we are supported by MITI, we are obliged to work on behalf of Japanese companies.

'Do you feel that the West has changed its attitude towards basic research?'

In the US, there used to be more 'will' to do purely basic research; they were trying to lead the world. But they seem to be forgetting this. Young people are not too bad, but the management class has become so demanding. They look at research purely in terms of profit. If they are pursuing a policy similar to the 'trinity' idea I have described to promote the application of pure science, I think they are making a mistake.

'Would you disclose all the information accumulated at ICOT for the benefit of the world?'

I have a clear point of view on this. When we started this project, there was much discussion about making the results of our research available, not only within industry but also among government officials. I demanded we should enhance the international community by disclosing all our results free of charge. I was attacked for this, but at the time there were a few independent minds inside the government, so we couldn't be stopped. We have made everything open during the whole ten years of the project. We have been asked why we released certain results without permission, and we have been told to wait for a while before making some software available. But in principle, I have tried to disclose everything free of charge. I am convinced that all governmental projects should be in the public domain. Although this is not yet completely the case, I hope the world is moving towards it.

'Could you describe some technical aspects of the Fifth Generation Project?'

The Fifth Generation Project, as I told you, has a very simple theory behind it. This is the structural concept of 'parallelism' and the methodological concept of predicate logic. This predicate logic is the core of the concept of 'inference'. Thus, we believe that the next generation computer is the parallel inference machine.

'Today parallelism has an established position but this was not true ten years ago.'

The majority was against it. Parallelism became established through the commitment of ICOT.

'What is the motive behind PIM?'

Let me start from the basics. Between hardware and software, lies the level of a machine language. The machine language issues instructions such as, 'put this data here' or 'bring it from there'. It is a matter of the recognition of either plus or minus. From a mechanical point of view, simple operations stand for orders, and by combining these orders, the computer can carry out different sorts of tasks. This combination facility is software. Those who design hardware make various devices which enable these orders to be executed as quickly and cheaply as possible. However, the software has to sort out a division of labour with the machine language or establish a border with it.

'So you mean there is a definite hard-soft border in present generation computers.'

Yes. But in present generation computers, this border or break works to the hardware's advantage. Machine languages are all too simple. So now, various burdens have been loaded on to the software. We must reconsider this situation, and try to redraw the borderline at a higher level.

'So the first aim is to change this border between the two worlds. It must be an absolute requirement for the future development of computer technology.'

Yes, in other words, the technological developments have favoured the hardware more than the software. Of course there are difficulties for those people actually developing devices, but it is a simple aim to make them quick and less expensive, and quite easy as a development principle. What used to be a huge box or even the size of a room now turns out to be a single chip. On the other hand, if development of the hardware progresses smoothly, it may have power to spare. So we should shift the burden on to the hardware and lighten it on the software, and then the latter will be of much higher quality, and be much more to the user's advantage. That's our basic philosophy.

'In order to establish this new border, your project dealt with both hardware and software?'

Yes, we dealt with both. The key idea was, I repeat, Parallel Inference. The line was drawn hypothetically as a research project at a higher level than in current computers. In order to do this, various new methods have been tried. We are actually trying them out on an experimental basis and constantly reevaluating. We have tried to further develop the software on its own side of the border. We have therefore formed two teams. The project is not complete until each team's experience has been transferred to the other, so interaction is important.

'Up to what point has the hardware been developed?'

There are two steps; one is to develop the devices and the other is to assemble them into a chip which fits into a box. In our project, we are assuming that the devices can be based on silicon. We have presumed that our prototype machine will be established around the beginning of the twenty-first century, so we thought the devices would develop smoothly if based on silicon. Of course it would be even better if we made more progress. There has to be a new stage created by the employment of future devices, but at this point, progress is constrained by conventional technology.

'So you see Parallel Inference as the main architecture for computers which are based on these future devices?'

Yes, it will be one of the pillars.

'Have any actual models of the Parallel Inference Machine already been constructed in ICOT?'

We have experimental machines. I think a very workable one will come out within a few years. However, as I have said, parallel inference is a concept which applies both to hardware and software. The essential point is not just lining up the chips but connecting and controlling them. In this sense, the software is the central problem. Not only ICOT but many other groups around the world are trying to solve it.

'Including industry?'

They have moved inevitably towards the parallel inference machine. As everyone knows, the main factor in the crisis at IBM is the fact that the mainframe is being replaced by the workstation. The two main factors in higher machines to achieve this are the parallel machine and the neural computer.

'In neural computers, it seems that what used to be mainly software has changed into manageable devices, such as neural devices.'

Yes, I suppose the demand for devices is greater with neural computers. There are many 'schools' in our field, but our aim for Parallel Inference is to lessen the demand for hardware.

'Could you explain a little about another concept connected with the Fifth Generation Computer, the inference machine using predicate logic?'

Predicate logic, or first order predicate logic, aims to reduce those things which comprise the world to discrete points, and then to describe the world as the relationship between those points. ICOT's idea is that by using predicate logic, we unite the most basic level of machine language with a higher level of computer languages.

'Has this been successful?'

We created a machine language called 'KLII' which is based on predicate logic, and which succeeded in operating its machine.

'How about with the computer language of a higher order?'

KLII is a machine language, and naturally it cannot be observed by its users. Then came the question of whether predicate logic can be applied to the user interface. This turned out to be the main theme of the latter half of my stay at ICOT. The predicate logic that I described is quite a philosophical idea; it is also so from a mathematician's viewpoint. An engineering concept is essentially 'object-oriented', and is the main concern in current software technology. Thus at ICOT, we created a new language which develops predicate logic in this direction.

'Did you give it a name such as "FORTRAN" or "C"?'

We called it 'Quixote'.

'"Quixote"?'

This is a clever way of saying Don Qui Haute; originally it was simply called DON (Deductive Objective Nucleus).

'What was the worldwide reaction?'

Many people found it interesting. However, the actual compiler is still slow, and its power depends on various factors. It needs to use homemade hardware, so it cannot be applied to every computer. But we are trying to make it applicable even for UNIX workstations.

'So what is the next step?'

Within the next two years, we will try to accomplish the unification of hardware and software not just at ICOT, but so that it can be tried all over the world. If our Parallel Inference Machine concept is accepted, it would mean our experiments have been successful.

'Could you tell us what will happen after the Fifth Generation project. Will the further development of computers continue into the sixth and seventh generations?'

I have a different view of it. The reason I am working on the fifth generation is not because it is the next one after the fourth. There is a great river of technology which has its source in von Neumann's invention, which then passed through its second, third, and fourth generations. This river will go on flowing. But the first generation was the greatest breakthrough. Presuming that the fifth generation computer succeeds, the sixth, seventh, and the eighth will only be further developments of the fifth.

'Do you think then that the fifth generation is the second most important breakthrough in computers since the first generation?'

I hope so. Of course they will also progress into other areas, but the main channel of progress will be through the sixth, seventh, and eighth generations, which will flow out of the fifth generation project.

'So the next breakthrough will happen only when a new theory supplants the one behind the fifth generation computer.'

Yes, even the fifth will be obsolete within forty to fifty years. There is always another invention on the new-technological horizon. Maybe that will be the ninth generation. I prefer to think of it that way. Neural computers have various possibilities, but I feel that it will be another forty or fifty years before they become computers in any real sense, though there are people who are much more optimistic than I am. Of course their technology will develop gradually and some things will be of partial use, but all the same I think it will take that long to effectively replace the current architecture. Therefore, it seems that although these large projects do occur from time to time, we will have to progress also in a more piecemeal way.

'So the fifth generation project has been in existence for ten years, funded from a large budget, but the sixth may be handled differently?'

For example, it won't be advisable in the development of neural computers to adopt the same pattern as for previous ones. For neural computers to be actually realized, we have to look forward to the equivalent of the ninth generation. But, unless we take stock every five to ten years, development will break down.

'So you think the image of the computer itself is going to change with the post-fifth generation computers?'

Current computers can be defined in many ways, but their main feature is their capacity for calculation. What we will do next is to get computers to make decisions, that is logical inference. Instead of the usual arithmetical operations, they will have to deal with 'meaning'. The human brain is expert at this while the conventional computer is a mere amateur. But this does not necessarily mean that they will be able to infer in the same way as the human brain does.

'So you think after the fifth generation, computers will at least be able to infer logically even if they will not be able to know the real meaning of such inference?'

Yes. There have been many fantastic proposals for AI; the computer as a thinking machine. But the way I see it, the next task for computers after calculation will be logical inference. Of course, many conventional elements will remain. In our research programme, we have to study the next thing which the computer might reasonably take over from the brain, which is I think logical reasoning.

'You consider that rather than making computers calculate at even more enormous speeds, we should develop a computer which is one step closer to the brain. There are many people who are looking at the possibility of simulating more intuitive functions of the brain than just logical reasoning, for example pattern recognition. What do you think of this?'

This is only one option for the development of computers.

'In which direction would you like to go?'

Pattern recognition, whether or not it imitates the human brain, remains a very important technological challenge. There is a need for it too. However, for myself, what I'm most interested in besides logical inference, is voice recognition.

'For example, the recognition of an individual voice out of a confusion of noise (the "party effect") is a difficult problem.'

As far as I know, there isn't anybody who is seriously looking at the problem. But we should try even if it does seem too difficult.

'Isn't it just like adjusting to a certain frequency band and then isolating the voice after filtering.'

And yet no one is really trying to do it. There is a consensus of opinion that voice recognition cannot be achieved without addressing the problem of meaning. But is this really the case? I feel there are many aspects which could be solved by sound analysis alone, and I don't think we should pass over this stage so easily. I will address this problem now I have left ICOT, to be continued at my leisure. Maybe we shouldn't talk too much about this.

'Science and technology are rapidly advancing in your field, with fuzzy logic and neural networks becoming the fashion recently. What principle of research have you adopted, as you have not been trying to follow the fashions?'

Computer technology tends to be a synthetic technology. In that sense, to stay in a leading position, researchers should try to synthesize everything available into a novel structure which is different from the previous one. I suppose this could be said of other areas, for example, engineering, but to progress from what has already been established, and invent a structure or architecture which puts these elements together seems to be the most advanced kind of science. This is especially true in computer technology.

'So you mean that the fifth generation computer will come about by synthesizing concepts from all the available technologies?'

Yes, we aren't waiting for something to fall out of the sky; we are developing the technology of earlier computers so that something new is gradually brought into being. By adding new concepts we are trying to develop a new shape.

'You said a little earlier that, in the future, computers will be roughly divided into two types: those which can make enormous calculations, and those which deal with meaning in imitation of the human brain. I would like to know whether you have a dream of how such computers can be matched together with human beings. How will computers change the world?'

If I could say one thing first; from the research point of view, there are so many ways in which we could fill the gap between human beings and computers. An example in point is brain physiology, which is a valuable discipline in itself, but there have been few cases where computers have been used to help bridge the gap. Even now, it is very large. But from now on, I hope the situation will change. Look at how far both disciplines have developed even since we were students! There will definitely come a time when computers are more flexible and when we have a better understanding of the brain, but we don't know if this will take thirty or a thousand years! So I think more scientists should try to explore the ways the brain functions.

'Can we concentrate on the near future, somewhere in the early part of the next century?'

In a broad sense, the pattern of life will be different from what our generation can imagine; it might be better understood by the younger generation growing up with comptuer games. There are things that have already appeared on the market; current computers have become quite compact, and if it develops further, even school children will be able to take their computers to school. Microchip technology will progress more and more, and I wonder what will happen when the computer is really a part of our daily life.

'Do you think that people will be degraded by this progress?'

This may happen for a while.

'For example, if computers and robots are doing most of the work?'

In the long term, I think intellectual work will be regarded as a sport. Intellectual labour is admired now, but gradually, it will become a sport, and play no part in the day to day workings of an office. Then we will feel that we don't have to do it, and we will avoid it. There may be those who will actually lie down on the job, but I guess this will be only for a transitional period, a phenomenon which has occurred in the past with other technologies. For instance, our legs were weakened by the invention of cars, but after a while, people started jogging to make up for it. We adjust as a matter of course and, as a result turn work into sport.

'None of our workers calculate using a *soroban*[2] any more; the actual calculation is done by the computers, and yet *soroban* schools still exist.'

I'm not sure if it can be regarded in the same way as chess, but all the same it can be interesting for its own sake. Sport can be enjoyed by everyone, whether professional or amateur. The leading role is taken by the professionals. In intellectual work, the most expert jobs are beyond computers now, but eventually some of them might be done by expert systems. Actually ICOT produced a legal reasoning system which may become a 'computer judge' in the future! Thus most of our intellectual activities will gradually turn into sports.

'And what will remain in the end as truly intellectual work?'

Pure research I think.

'And perhaps truly creative work?'

Those who have physical strength and well-developed motor systems will become good golfers or baseball players. In the same way those who are interested in intellectual things and who have sufficient talent may become creative researchers, and so everyone in the twenty-first or second centuries will become either a sportsman, or an artist, or a researcher.

'Do you think the computer will eventually be superior to the human brain?'

There are some ways in which it will, although maybe not in areas like intuition and emotion.

'How about pattern and voice recognition?'

Pattern recognition still needs more research. It is quite easy for human beings but eventually most of its aspects will be technically overcome in computers. But many difficult problems remain. There is for example the speed with which ideas come into the brain. The machine which we call the human brain is, from a technological point of view, an ultra-parallel inference machine with outstanding functions, and I suppose this has a lot to do with the way the mind flashes out ideas. This will be the last part to be solved; at the moment it is little more than a dream.

'The brain, though it has some digital characteristics, is mainly analog in character. The brain is such a dynamic thing that, for example, how well it works is affected by how well the person has slept.'

Yes, maybe this is a characteristic of living things. In this respect, conventional technology has more merit, because it is tougher. For example, a

[2] An Asian tool with sticks and balls to calculate.

silicon semiconductor is both unbreakable and steady in its functions. However, there are people who are trying to develop 'soft' computers, so there is a possibility of 'sleepy' computers being developed.

'Scarlet O'Hara in *Gone with the Wind* defers thinking about a problem until the following day. Human beings are always doing such things. That is, depending on one's physical condition, the brain takes precautions to prevent confusion and defers action until later, after it recovers. Humans can do such things because they are alive, but how about computers? Will it be possible to put these human characteristics into computers?'

I'm not sure, but activation of the computer's structure will be one object of research. What I sometimes find interesting is the division of consciousness and unconsciousness in humans. It may not be quite the same as in Scarlet O'Hara's case, but sometimes when one gets stuck and stops thinking about a really difficult problem, it may happen that the solution suddenly flashes into one's mind the following morning. We often have such experiences even if they may not be the insights of genius. There must be something in the unconscious which is added to conscious thoughts and sends something flashing to the surface. I now wonder about the possibility of an 'unconscious computer'.

'So the theme for the tenth generation computer may now have been decided!'

Amari

SIX

PRINCIPLES GOVERNING THE BRAIN AND COMPUTERS

SHUN-ICHI AMARI

AFTER ELECTRONIC computers became indispensable elements in civilized life in the 1980s, there began a movement towards making them much closer to human brains. One of the most promising of these attempts was the emulation of the neural architecture itself by computers, so-called 'neural computers'. The theory behind neural computers has not been established yet, but the most advanced idea at present is generally called 'neural network theory'.

Professor Amari is one of the first-generation graduates of the mathematical engineering course of the Department of Engineering at the University of Tokyo. He is the founder of a new field of information science, 'differential geometry of information'. His approach has been characterized by its high level of abstraction. Thus, it was not as an insider in computer science that he saw the problem of today's computers, but as an outsider. This unique approach has enabled him to become the front runner in neural network theory. Conceived during the 1960s as a purely mathematical model for the processing of information, it has now become the mainstream approach in the endeavour to transform current computers into real thinking machines.

In this interview, Professor Amari focuses on the relationship between the computer and the brain from his own unique and intriguing viewpoint.

———

'Research in neurocomputers expanded greatly around 1986 and seems now to be past its peak, but I understand that the idea of the neurocomputer itself is by no means new?'

Yes, since Turing devised the computer, the brain has been a sort of rival, or, if you like, an overshadowing presence, something to be emulated. Von Neumann, the architect of the modern computer, was also fascinated by the brain. He spent a great deal of time debating whether or not the brain worked in the same way as a computer. It's interesting that while the computer can be described perfectly as a Turing machine, the brain, although we live with it and have an intuitive grasp of its workings, has so far revealed nothing about the principles of its own operation.

'How old were you when these arguments about the brain versus the computer were going on in the West?'

I was only a student.

'Were you interested in arguments like Von Neumann's at the time?'

Not very much. Actually I was so inclined towards mathematical ideas, I was not even interested in computers! It has to be said that computers themselves were not very fancy at that time: when I was a student there was not even a programmable computer in Japan.

'When did people first begin to study the brain with a view to the development of computers?'

Well, the first attempts to emulate the brain were made in the early sixties when I was a graduate student. This was called the 'perceptron'. But they ran into all kinds of difficulties when it came to reproducing the learning process, which is the most important faculty. A computer program is written by a human being, who carefully plans out the method of operation from start to finish, expressed as a series of commands. This is a problem when you attempt to change the program by employing a learning process. The brain, on the other hand, may not know how to do arithmetic, but if given an example, it is able to learn how to do it somehow or other.

People began to ask whether it was possible to imitate this particular function of the brain, and this led to the invention of the perceptron. The perceptron is a pattern recognition device, which uses the learning process to improve its own ability to recognize. Naturally, it attracted a great deal of attention. Computer engineers and electrical engineers flocked in to the field and soon the first neuroprogram was developed. It captured the imagination of engineers all around the world, but the boom didn't last beyond the sixties.

'Why was that?'

You could call it the 'technology barrier'. To put it simply, there was no hope of building such a device with the technology available at that time. So people began to look for ways of learning from the brain without imitating it directly.

'Artificial intelligence?'

That's right. Artificial intelligence was the main line of research in this field in the seventies. But actually, the field of neuroscience in those days was so little advanced that no one really had any idea how to begin learning from the brain.

'I understand you entered the field of neurocomputing after the perceptron had already come and gone for the first time, and that you are in fact credited with its revival?'

I became involved with neural networks in the late sixties and into the seventies, after most people had abandoned neurocomputing.

'You said you were not interested in computers when you were young. Why did you become involved with them later on?'

My background was in mathematical engineering, in which one attempts to gain understanding of complex real-life phenomena by mathematical methods. I wondered whether it would be possible, instead of struggling to understand the phenomena themselves, to step back and take a mathematical approach, that is to try as far as possible to put things into some kind of mathematical order. This, I felt, might lead to new discoveries. Before I became interested in pattern recognition and learning processes, I worked on graph theory and continuum mechanics using topology and differential geometry.

'So, you decided to stand back from both the brain and the computer and take up the challenge from a different angle. Could you give me some examples which would demonstrate your approach?'

Imagine those imperfections inside a piece of metal which we call dislocations. In the science of materials, each of these dislocations is considered to be a separate imperfection. However, our approach identifies these dislocations not as individual entities but as something distributed in a continuous pattern. If you apply a force at a given point, the dislocations move and vanish in pairs. We attempt to analyse the dynamics of this in geometrical terms. The material forms an abstract space which is not Euclidian but curved. The principle of relativity states that the curvature of space-time is manifested as gravitation, and occurs because of the presence of matter. On quite a different scale from relativity, in the structure of a material with defects such as we have described, it is these defects that create the field of distortion.

'So you tried to introduce this kind of abstract approach to research into neural networks?'

Yes, using this kind of methodology, we attempted to sort out once and for all, in a mathematical way, the complexities and mysteries of the brain, and its vast numbers of elements. The cerebrum has a hundred billion neurons: when one neuron is excited, it affects others surrounding it, and this effect spreads. Interactions occur one after the other, linked in temporal and causal relationships. We call these interactions 'dynamics'.

'You start then from principles, and not from individual occurrences. How do you proceed from there?'

Let's consider possible mathematical models for the kinds of phenomena, for example the dynamics which occur in systems such as the brain where a vast number of elements are interlinked into natural patterns. Rather than arguing in detail that because such and such is true, then the brain must be so, wouldn't it be better to suppose that there is some kind of principle of dynamics which operates when a large number of elements interact in natural patterns, and that the brain may work on these principles too? I first looked at the brain by using this kind of abstract approach. By examining the learning process and data processing in this way, I hoped to glean some idea of the mechanisms of the brain, after which one could look into the ways it actually functions. This was the mathematical engineering approach I adopted, taking one step back from the problem itself.

'So you initiated this different approach to neurocomputation, after the first wave of enthusiasm for it had waned in the seventies?'

Yes, it was the kind of work I was absorbed in during the seventies. In those days, the brain was being studied from many angles all over the world, especially in biomedicine. But because the biomedical approach tends to stick with the brain itself, biologists tend to focus on the material out of which the brain is made and their micromechanisms, and never seemed to get around to the question of how the brain handles data processing and other such abstract functions. This was the ultimate goal but it always seemed far into the future; then groups of people emerged all over the world who were taking the kind of approach I've just described, devising engineering models and seeing how these models worked. Now we are seeing a boom and there are several thousand of us, but at that time there were perhaps ten people in the whole world.

'So you approached the problems of neural networks from a mathematician's viewpoint, or rather that of a mathematical engineer, but using computer engineering and brain physiology as a basis for your research? Could you describe this in more detail?'

Suppose you take the function of the neuron as it is understood in biological terms and link a number of neurons in random circuits; then study what they can and can't do, mathematically. Theoretically, all you have to do is write a number of equations to explain mathematically the nature of the brain, but there are too many degrees of freedom to allow an exact solution. And even if you arrive at a solution it will be too complicated to gain a meaningful understanding of how the brain actually works. So you have to simplify the model to a point where these equations could be solved in a meaningful way. When we were ready to make the model a little more complex, we used computer simulation. But the real work was in establishing the basics using a pencil and paper, and it was not until the end that we used a few computer simulations.

'So you start with a very simple model and then proceed step by step to more complex models.'

Exactly. We began with random circuits, then looked at neurons arranged spatially in layers, such as the cerebral cortex, and then looked at the kinetics of the pattern by which the excitement of one neuron is transmitted to others around it. Then, by way of basic research on the learning process, we investigated possible models for memory where the links between neurons change as a result of this action. We started from basics, and added a learning process and a field, step by step. That was the main thrust of our work during the seventies.

'And this pioneering work was followed by the "neuro-renaissance" in the eighties?'

In 1982, about a year before the neural computer boom started in America, there was a US-Japanese seminar on neural networks. This seminar was sponsored by the JSPS[1] and the National Science Foundation of America; in it there were symposia on set themes and opportunities for the exchange of ideas. The Japanese attendance consisted of engineers such as myself, biologists, and mathematicians; but American biologists at that time were not interested in such models, so they had nothing to say. Anyway, many American and European scientists attended, and over the next year, things started to change in the US.

'Why was that?'

Well, there were a number of reasons for the boom, but one particularly important factor was cognitive science that is, if you like, the study of the mind. You can describe the workings of the mind in literature, but it's quite another matter to try to explain them in mathematical terms as a logical system. Of course, literary explanations lack scientific objectivity. Cognitive scientists relied to a great extent on computers, such as they were in those days, and tried to use them to construct models of the mind. For instance, by inputting information describing a given situation, having created in advance a program that would give the answer 'a human being would do such and

---

[1] Japanese Society for Promotion of Science.

such in this situation', this program would be objective, and by studying its findings you would get a picture of the workings of the mind.

'So this was the heyday of AI!'

Yes, this program would be virtually a form of artificial intelligence; and so cognitive scientists were attempting to study the mind by writing these programs in collaboration with AI researchers.

'I think it is true that the AI-cognitive science research program soon began to display serious limitations, right?'

It soon ran into problems. If one were to focus only on the serial-data processing side of the mind—in other words logical thinking—it might be possible to construct a mathematical description of that process. However if, for example, one considers all the errors that the mind produces, one can't help feeling that these are not just mistakes of logic but also the results of the interactions between elements which take place as the brain processes data in parallel. These parallel interactions allow the brain the process data rapidly but they also leave room for mistakes.

For this reason, it was understood that if we were to solve the mysteries of the brain—rather than relying on serial processing programs alone—it would be necessary to devise systems which realize information processing by interacting between connected elements. This concept was called 'connectionism'.

'So, you are saying that the AI-cognitive science research program merged with the neural network approach?'

There is another source for this approach which I would like to remark on because it would be of important testimony in the history of science. This was another clever idea, produced by physicists working with a simple material called spin glass; they applied the dynamics of spin glass to the study of neurons.

These two trends converged with our work on neural networks and provided the basis of the main trend of research in the eighties.

'So what is the main factor behind the current boom in neural computers?'

As very often happens in the history of technology, progress on the hardware side was the main factor. In the sixties, if you devised anything even remotely complicated, your chances of making it happen, technologically speaking, were about nil. But now people are talking about making 'neurochips' using LSI (large-scale integration) and optical elements, and even if you are not trying to make anything as complicated as a neurochip, silicon LSI technology will allow you to build parallel junctions and other complicated systems.

So we have come to the point where, if very complicated systems with many interlinked and interacting elements are needed for progress, the technology will soon be available to build them. For instance, in a situation where it would be difficult to make a junction, it will soon be possible to take advantage of optoelectronics technology and build an optical junction. Of course, no such miracle has yet happened, but judging from the progress of technology, I feel sure that it's not that far out of our reach. With this kind of technological backup, everyone is keen to have a go.

'So that's the hardware side. What factors are there on the software side?'

One factor is that there has been a certain amount of despair about artificial intelligence. During the eighties, artificial intelligence became popular a little earlier than neuroelectronics, and once the boom was in full swing, too many people went around saying that artificial intelligence could do anything. Of course, simple tasks can certainly be solved by the logical processing of artificial intelligence, and the software and systems for tackling them were developed, but in practical terms, it is very difficult to write a program that takes account of all the possibilities that arise. Once you start dealing with real life, you begin to find contradictory data or data that is vague or incomplete. To design a program that can cope, taking account of all the possibilities, is pretty well impossible, and then if all the information can be put into a database you will find that it's difficult to make the machine learn; and there are all sorts of other problems too.

For reasons like these, 'neuro' became not much more than a trendy commercial expression during the eighties and on into the nineties.

'Yet at the moment, we have software which uses neural networks even on PCs!'

I think that practical applications have come surprisingly early. For instance, apparently stockbroking firms are using them to predict share movements, though the systems used are secret and, I'm afraid, beyond my understanding! One thing is certain: all practical applications of neural networks so far are commercial and do little more than scratch the surface, though I would agree that they do prove the usefulness of neural-style computation. It will probably be another ten or twenty years before a really serious neural network machine is developed.

'So neural networks are still at a very difficult stage?'

I believe we haven't even reached the stage where the field can be dismissed as merely 'difficult'. I think we are still at the 'nebulous' or 'fuzzy' stage.

'All in all, do you believe it is possible in neural computing to "learn from the brain"?'

My ideas probably differ from other people's. Basically, you can describe a modern computer as a Turing machine, and its design concepts as being of the von Neumann type. They are all 'machines for the processing of symbols'. Addition is the simplest example of symbol processing. Any word, expressed logically or consciously, will be processed electronically in a modern computer as a 'symbol'.

But, take this *chawan*[2] for instance. Suppose the modern computer is presented with a *chawan*: does it actually process the bowl using the symbol for *chawan* or 'cup'? Or doesn't it rather understand and process the cup in terms of its form? The most important function of the human brain, I think, is that it doesn't process an object in terms of symbols but intuitively and unconsciously as form. Also many items related to but not actually represented in symbols—like the raw material and even the history of this *chawan*—are activated at the same time as the brain grasps the *chawan* as a form.

'So basically, you don't agree with the idea that the brain is processing information using symbolic, or digital, representation.'

Of course the brain is doing that. But symbol processing is not the only job it is doing nor is the most important or fundamental.

'The "form" approach you mentioned is the most important then?'

Yes, I think so. And also I would conjecture that this method of processing data by 'form' rather than symbol conversion is founded on some abstract principle of which the modern human brain is simply one biological example. The human brain is after all the product of an extremely random evolutionary process and it is unthinkable that no other comparable 'soft' data processing machine should exist.

'So as we don't know whether the human brain is the most advanced possible device of this sort, shouldn't we simply be trying to imitate it then?'

Well, maybe, but don't you think it would be rather unpleasant if we produced something that was too good an imitation! It would also have all the follies of the human mind! As a rule, technology does not imitate nature and living things but merely tries to pick up hints from nature. The aeroplane takes hints from the bird; it does not try to imitate it. If man had tried only to imitate legs, he would never have invented the wheel, or the automobile for that matter. So in data processing...

[2] Japanese tea cup.

'…the computer is to the brain what the automobile is to the legs.'

That's right. If you see the human brain merely as a symbol processing machine, it is no doubt more advanced than the brain of a dog or a chimpanzee, but it comes nowhere near a computer.

'So we can ask the computer to take over our symbol processing functions in the same way as we rely on a washing machine, which is much better than a human being at cleaning dirty clothes!'

The brain does have some of the characteristics of a symbol processing machine, but it is actually better at flexible data processing. But, I would repeat, maybe its workings are merely an example of more abstract principles. In which case, if we could grasp this non-symbolic principle of data processing, of which the brain is just one example, it might enlighten our actual research into the brain, and then again we might also be able to put this principle into practice technologically, and build a machine which is quite different from the brain.

'In other words, it is possible that there might be quite a lot of different mechanisms that could work out of this principle?'

It may now be possible to build a 'non-brain' machine that thinks in patterns and learns in flexible ways. This could even be done by making von Neumann machines in parallel. In fact, there are already simulations using current LSI technology. But in the distant future it seems likely that we will have to abandon electrons and use light instead if we are to build systems which have the complex and dynamic connections of neural networks.

'So you are optimistic about the future of hardware?'

In some senses I am. However, please don't think I am saying that optical technology for example can bring about the ultimate neural computer. The most important thing is to create the software to develop methods of data representation, to understand how to do the computations, etc. We still have a lot to do.

'How do you see the brain in terms of hardware?'

I can only look at research on the brain as an outsider, but over the last twenty years I believe significant progress has been made towards understanding the mechanism of brain function. Sophisticated and creative intelligence is still a distant goal, but I believe we have already found the first clues to understanding perception and memory. At the molecular biology level, and at its interface with cognitive science, many people are working with models of neural networks.

'So you think neural models are relevant to understanding the real brain?'

Yes I do. We ask questions about the brain such as: what are the preconditions for building intelligence into the network of neurons which comprise the living brain? How many neurons do you need to handle a given amount of information? What percentage of these neurons should work to provide the optimum conditions for memory? It will soon be possible to predict the answers to all these questions using neural network theory and through experimentation.

'Let me take an example. Recently there has been a great deal of discussion about the column structure of the neurons related to vision, that is groups of neurons which react specifically, for instance to the shape of a pair of lips. What principle is at work here?'

That's an extremely interesting question. I believe that this is an example of the common phenomenon of 'system self-organization'. Most people think of learning as a down-to-earth process which evolves through a process of trial-and-error. However, for a complex system such as the brain, it is inefficient to process all these individual decisions in turn. It is far better to establish some kind of arbitrary system, regardless of the outcome of the decisions.

To us, a rainbow appears to be seven colours. In fact it is a continuous spectrum and the seven discrete colours only exist in the mind. The reason that patches, columns, and clusters of neurons exist may be merely because the complex system of the brain needs to organize itself in some way.

'That's an interesting viewpoint.'

I might also suggest that the symbolic representation of information itself emerged as an extremely advanced form of self-organization of the nervous system.

'It seems to me that you have broadened the concept of "information" and think of it in quite abstract terms.'

But actually, my real life-work may not be in neurocomputers. In twenty years' time I don't expect to be remembered as a pioneer of neural networks.

'So what do you feel is your main achievement?'

I hope to be remembered in connection with the new field of 'geometry of information'.

'Can you describe the theory?'

It involves, for example, introducing the concept of duality into Riemannian space. There are only about twenty people in the world who really understand this field, though their number is ever increasing, and many physicists and information scientists are going into it. In very simple terms, the idea is that computers see information only in terms of a series of individual 1/0 signals; but in information geometry, we think of the totality of information as forming one abstract space.

'What sort of applications would there be for this?'

Well, take statistics for example. It's a very old subject, but one in which the basics don't seem to be very firmly established. Information geometry tries to redefine these fundamentals from a different viewpoint. It also helps to shed new light on control theory.

'So can neural networks also be constructed according to the principles of information geometry?'

Well, initially, I thought the two fields were completely unrelated, but later I discovered that the Bolzmann machine—one type of neurocomputer—is actually described exceedingly well by information geometry theory.

'Your "abstract and mathematical" approach to engineering is something that developed after the war, I believe?'

The predecessor of mathematical engineering was known as 'applied mathematics'. Most people said that this discipline was for those people who couldn't manage 'pure' mathematics. This view is still predominant in Europe, but the belief that there is pure mathematics on the one hand and that mathematical engineering is an application of it is quite wrong.

'In America, this field was developed by exiled scientists like Wiener and von Neumann.'

That's right, and its main exponent in Britain was Turing. Anyway, it is interesting that their ideas were developed in America.

'And what about Japan?'

At the University of Tokyo, during the war the most prestigious area was the aeroengineering department, and the best mathematical minds gathered there. After the war, the department of aeroengineering was changed into a mathematical engineering department, and so the mathematical approach to engineering developed in parallel in Japan and America.

'In Europe, as I understand it, engineering is considered a second rate discipline.'

Yes, I think it is. But this is not the case in Japan. Science in Japan is effectively science and technology; we don't have the conventional division that exists in Europe between pure and applied science.

'So why is there this difference between America and Japan?'

We don't have the American spirit of individual adventure in Japan. Japan is basically an egalitarian, horizontally structured society and it is difficult to do anything really pioneering or adventurous here. But this is changing with economic development.

'How were you able to overcome these obstacles?'

Well, I was especially lucky. And because I was not particularly concerned with money, I was able to do as I pleased.

'Why weren't you interested in money?'

I am not saying I dislike money! When I started my research career, Japan was quite a poor country. With mathematical engineering, all you need is your own neural networks, which is very cheap.

'As regards academic freedom, do you feel Japan is perhaps closer to Europe than to the USA?'

Probably for professors, yes. In Japan, however, there are virtually no restrictions on graduate students. Academic freedom is fully entertained by students, too! That is the big difference between Japan and the West.

'Would you like to live in America?'

I was offered a post there once, but I turned it down. In America, you spend most of your time on publicity and fund-raising. I couldn't stand that. In Japan and Europe we have a much easier time of it. I think it would be very sad if Britain was being Americanized.

'If you hadn't become a mathematical engineer, what would you like to be?'

Probably a GO[3] player, or maybe a RAKUGO[4] story-teller.

---

[3] Oriental game of geometry.
[4] Traditionally popular comic stories.

'And what do you think of today's young Japanese people?'

They are completely spineless. They haven't even got any team spirit. I'm not saying that it was a good or bad thing, but in the past the student movement was directed against the establishment. Now that kind of thing is unheard of. I feel it is important to fight the establishment to a certain degree.

'I suppose in any society a spirit of questioning authority is important. You have to challenge established theories, otherwise you will never invent anything new.'

When you are a student at university, you have to tell yourself that your professor is past it and that he's not much good. But it's equally important for the professor to insist that he's still better than his students. Competition is good both for students and their professors.

'But can't we then place our hopes in today's younger people?'

When I think of my graduate days, I realize that I felt antagonistic towards my professor, but in fact he was really very much on-the-ball. He was still reading and studying. It made me grind my teeth to think that someone who was so knowledgeable could still study as if his life depended on it. Yes, I really feel it's important to have a spirit of rebellion.

'Do you have a message for young people?'

Yes. Join me in the search for those higher principles that unite the brain and the computer!

'Do you see yourself as a leader in this search?'

No, maybe just a cheerleader—and, I would add, don't follow your leader blindly!

Yanagida

SEVEN

BEYOND INTELLIGENT MATERIALS

===============================

HIROAKI YANAGIDA

THE HISTORY of civilization has been characterized by the materials on which its periods were based: the stone age, the bronze age, and the iron age. From this viewpoint, the last few hundred years can be seen as the period when the iron age reached its final phase, the age of steel, characterized by a daunting quantity of machines, cars, and skyscrapers which are the essence of the modern city. Although this industrialization is seen to have come to its limit, there have been few proposals about what material civilization will be founded on in the future and what philosophy will provide its basis.

Professor Yanagida of the University of Tokyo is one of the leading scientists in the study of ceramics, which may well be the key materials in the civilization of the future. He was the first to point out that ceramics can become 'intelligent'. As this is now a design principle in engineering in a wide sense, it has led him into a new field of technological thinking. He is now the advocate of innovative principles of technology, such as the 'principles of simplicity', 'technodemocracy', and 'technology with virtue'. He is therefore not only leading the development of new materials for the future, but also trying to change the climate of modern technology in general. As his general views have been expressed at length in the introduction to this book, he focuses in this interview more on his own subject as an application of these new principles.

————

'Your research has mainly been on fine, that is advanced, ceramics. What are you focusing on now?'

I specialize in material science and the electronic properties of fine ceramics. Materials have been categorized as structural materials and functional materials. Up to now, I have been dealing with functional materials. Making the most of the durability of ceramics, I first tried to use it as an electronic device, for example by applying it in sensors.

'I believe that your research was at first mainly on basic structural materials and then progressed into functional materials, and then on to what you call "intelligent materials".'

The word 'progressed' makes it sound as if conventional materials are in some way inferior, so I would prefer to say that the new materials coexist with the conventional ones.

"Intelligent materials" is a concept you proposed as a new stage beyond previous classification systems such as structural versus functional materials.'

It's a word which was invented in around 1986. I should also say that it's one of many English phrases which have been coined in Japan. In the West where people tend to think that only human brains have intelligence or spirit, it may be difficult to accept the idea of materials having intelligence. The computer has come about as a result of an attempt to emulate the functions of the brain in isolation from the rest of the body. In the field of materials, people tended to think that because an inanimate thing cannot make judgements, a computer, or a chip should be attached to control it. Intelligence comes from outside; it is something to be added. My proposal rejected this idea of intelligence. I basically proposed that something without a brain can have intelligence.

'For example, the human body itself can perform very intelligent tasks unconsciously: the immune system can function without being taught by the brain. Is this the sort of thing you mean?'

Exactly.

'Then, under what conditions can materials be called "intelligent"?'

As you know, a living thing has various self-organized mechanisms which maintain its life activities; for instance, it has 'self-diagnosis' to judge its own condition, 'self-adjustment' to the environment based on that judgement, and finally 'self-recovery' in case it breaks down. I think these are the basic intelligent functions of self-organized systems like living organisms. The immune system you referred to has exactly these functions without being taught by the brain. We want materials to use these functions, so you could say what we are doing is quite an experiment.

'Do you mean that the ideal aim of an intelligent material is to be like a living thing?'

No, not exactly. Let me explain. Out of all materials, I have specialized in ceramics. The outstanding feature of ceramics is their durability. They are heatproof, corrosion-resistant, and wear-resistant. I want to add to them an element of intelligence like that of a living organism. What I'm aiming at is to imitate the characteristics of living things in materials which can be used in environments where no life could endure. Therefore, it's not my intention to make materials look like living things in their own environment.

'I often hear people say "you should learn from life", but eventually, we have to go beyond this. It's true that proteins and other organic materials which make up living organisms have their own flexibility, but they are also fragile and weak. We are too delicate to endure fire! We must try to overcome these weaknesses, mustn't we?'

We don't know if we can overcome them completely but it is certain that we can make something which is stronger.

'Could you explain why you thought you could make ceramics intelligent?'

Well, there are several reasons. But there's one point I would like to emphasize now. It is a challenge to the current tendency to give everything intelligence on the model of electronic computers. The way we rely on computers in science and technology nowadays seems excessive to me. I would say this is not so much a computer culture as a cult! For example, the more we feed back every judgement to the computer, the more we need electrical leads. I would even call this overwhelming situation a 'spaghetti syndrome' following the analogy of the crisis in modern medicine.

'You mean that advanced medicine makes patients in Intensive Care Units look like spaghetti, connected to enormous numbers of tubes?'

Yes, exactly. It's a problem if we rely too much on computers, because we tend to lose the original material when we put more and more layers of chips on it. That's why I thought something should be done inside the material itself, and not from the outside by adding chips. This creates a new point of view.

'Leads are the biggest problem when biocomputers and neurocomputers are put into practice. Does the same sort of thing happen in the world of materials?'

Yes, we tend to wire everything up and solve problems with computers. But if decisions can be made within the materials, we wouldn't need an electrical lead, and we would get less damage as a result of uproar in the electronic system. What's more, if we can let materials themselves have self-recovery or control, then I suppose the safety of the system will increase.

'A fashionable term for this would be "distributed systems".'

Yes, not central control but localization. However, we should not build up a distributed system by distributing computers! We have to help the material itself to become a system. My aim is to help the material have intelligence by making the most of its electronic, optical, and chemical characteristics.

'From a common sense point of view, it seems very difficult to endow materials with intelligence; ceramics for example have the inherent image of being something solid and cold. Would you give us a simple example of an intelligent material?'

From my own experience, I could describe the p/n contact humidity sensor which is made from p-type and n-type semiconductors. I think this sensor behaves very much like the human skin. Electric current flows at the contact points between the two types of semiconductor. Since they are not actually connected, there are places where there is no contact. When vapour enters the gaps between the semiconductors, it attaches itself around the contact areas and, as a result, the electric current flow changes. We can estimate the humidity by measuring the electric current. We do something similar with perforated semiconductor sensors, but the problem here is that they can't cope with a sudden change of humidity. The attached water doesn't go away; the humidity sensor needs to have its system reset; it is heated up in order to get rid of the vapour. With the p/n contact sensor you don't need this heating.

'Isn't it important for sensors to measure under the same conditions in order to be accurate?'

Not really. The old sensors judge what has been attached in the form of water and input this information to the computer in order to get rid of water. It then moves on to the next step. The p/n contact sensor measures the humidity by getting rid of water so that it can always be used in a fresh condition. This means that it waits for the next step while self recovering, which is much the same as what human skin does.

'Does it take advantage of changes in the environment?'

It measures energy by causing it to revert to its original condition.

'So what's the merit of it? Will it either increase the accuracy of measurement or its stability?'

Because it's a transient phenomenon, accuracy won't be that high. But on the other hand, there would be no hysteresis, that is to say, both the values measured coincide when the humidity is increasing or decreasing. In conventional sensors this is not the case. For this reason, ours will have good reproducibility, which leads to the maintenance of reliable information.

'Will it be able to obtain the information in a straightforward way?'

The way it measures humidity is like that of human sense. We feel ill when it's very humid, because we have to use up extra energy to get rid of water vapour from the skin. The sensor we made measures that energy. If we add air to it, the air helps to get rid of water, so only a small amount of energy is needed. It's like us feeling better when the wind blows. This sensor may not be accurate when measuring absolute humidity, but it is suitable for measuring effective humidity.

'So that's what you mean when you say it resembles human sense. What are the prospects for this sensor?'

Intelligence doesn't mean having a real brain. For instance, if a material is capable of self-diagnosis, it enables us to check the durability of a building whose walls are made out of it. At present, though, the only way to maintain safety is merely to make things 'strong'.

'The safety factor is often ten or even hundred times the necessary minimum, isn't it?'

If the material can let us know when and where it's going to break down, we only have to make the necessary changes at that point. Maintenance would be much more convenient. From now on, it will be important to find out how to maintain what has already been constructed. The importance of maintenance has been raised already, and I think one of the solutions to this is to let the material itself have a self-diagnosing function. Furthermore, it would be better if it could have a self-recovery function too so that it could recover when it begins to break: this is not easy though. The humidity sensor that I described includes this self-recovering element. Ordinary sensors couldn't cope with a sudden change of humidity, but we proposed a sensor that can register humidity while in the process of self-recovery. What's more, I've been thinking of what I call a 'stand-by mechanism'. With this, apart from self-diagnosis and self-recovery, a device would be on stand-by in its optimum condition. This mechanism has already in a sense been realized, for example, in the transistor. In this there is a gate between p–type and n–type semiconductors; if you allow an electric current to flow by reducing the bias voltage, it does so suddenly and consequently the electric current increases rapidly; this increase cannot be obtained without a stand-by mechanism, the bias voltage mechanism, so in this sense a transistor is very intelligent. If a sensor can have a stand-by mechanism, then we could devise a system which could receive its input under its own best conditions.

'Can you give an example?'

For this, we have concentrated on infra-red sensors. By using semiconductor barium titanate ceramics, we have succeeded in creating a sensor which maintains itself in its own best condition. I call this a stand-by characteristic.

'What about the other characteristics you mentioned?'

Self-adjustment, as a target for intelligence materials, is somewhat similar to self-recovery; it's a way of always returning to the best condition. A simple example is the adjustment of light through windows and sunglasses. As you know, in some sunglasses the colour of the lenses darken as they catch more light, and they also increase the amount transmitted when the light intensity lessens. The glass itself is doing this without any input of information from outside, so I would regard it as an example of a quite intelligent material.

'Self-adjustment could be defined as adapting to the environment without any input of information?'

Yes, what would be even better is something that can also adjust automatically to the amount of light, that could also radiate light under weak natural light conditions as well as darken in strong light conditions. If a room could be adjusted to a dark tone when one felt like staying in a calm environment, we could even plan a room which changed according to one's mood!

'Whether a material can self-diagnose would have a great influence on human life and on product value too. I hear you have proposed intelligent materials for construction, for example a mechanism which detects cracks in materials, where the only other way to check for them would be destructive testing.'

It is possible to do this even now if we ignore the economic cost. For example, there's a way to detect the signs of very early cracking electrically, but this system is very expensive. To make it less so, the only way would be to let the material itself identify these signs.

'Aeroplanes and automobiles have increased their speed; we'll soon be in the era of the space station. The most effective way to prevent accidents will depend on whether materials themselves can self-diagnose.'

The JAL[1] plane crash in 1986 was caused by worn material. The current method of checking for cracks in aeroplanes can only be done individually. Because I suppose I've proved self-recovery can be realized to a certain extent, my next aim now is to achieve self-diagnosis.

---

[1] Japan Air Lines.

'Could you explain this in detail?'

Well, CFGFRP[2] is something I investigated recently. It's a material which is used frequently to replace iron reinforcing bars in concrete. When concrete gets damaged by stress, fracture occurs first in the carbon fibre. However, glass fibre doesn't fracture yet because it resists the same degree of stress. So the condition of the carbon fibre changes when glass fibre can still structurally bear to the stress. The apparent electric resistance changes a good deal when fracture occurs in carbon fibre. As a result we get to know the stress on the concrete by checking the conductivity of the hybrid material, and can predict the danger of collapse long before it actually occurs. The older way to check for cracks is called acoustic emission (AE), a system which is very expensive. But it's quite easy if all you have to do is measure the electrical conduction.

'With less cost?'

Compared to acoustic emission, it is very cheap.

'While we are talking about utilizing changes in electrical characteristics, could a phenomenon such as weak emission of electromagnetic waves be included? I think there is a method of investigating the human body by observing the weak emission of heat.'

Yes, this option has been considered. If we use the weak heat emitted from the body for medical diagnosis, we can investigate the distribution of the temperature on the surface of the body. This is done with an infra-red sensor acting as a sort of monitor. I was surprised to discover that this system costs around Y10,000,000 (£62 500). Those systems which employ colour display are even more expensive. We are looking at how to make a simple sensor which is more reasonable in terms of cost. There is another sort of sensor which can detect infra-red rays from a stationary target. Earlier infra-red security systems could only detect small changes in rays when a person moved.

'You don't mean that there's no ray coming out of an immobile target, do you?'

No, although there are some, current sensors that cannot detect these rays when their intensity does not change. To change this, you have to chop the ray up and digitize it, to create the impression of a change, but to achieve this you would need extra electronics. We thought if the sensor could detect the infra-red ray itself, there would be no need for this additional chopper

[2] Carbon fibre and glass fibre reinforced plastics.

circuit. We have succeeded in working out how to do this using intelligent materials.

'Why couldn't earlier sensors detect these rays?'

A sensor that could detect a non–fluctuating ray was too expensive; less expensive ones couldn't detect non–fluctuating rays. The one we invented uses a semiconductor fibre. Making use of its characteristics whereby its resistance changes depending on temperature, it reports light intensity by changes in its resistance. A rough idea would be that by decreasing the number of circuits involved, we could say that it has become intelligent. A subtle change in resistance is quite easy to measure, because it's an electrical property.

'Your concept of intelligent materials seems to be that you add the unconscious flexibility of living organisms to inanimate materials which can endure conditions which no living things can bear.'

Yes, I'm thinking of a material that has the ability to self-diagnose, self-recover, and finally if necessary, break up and return to its physical environment. It would be a problem if it went on self-recovering and never broke! A good material should be a hard worker, and yet pop off at the end! There are some who think, especially in the US, that materials should become tougher and tougher, which is I think an ideal situation for building weapons. My concept of intelligent materials is that they can coexist peacefully with humans and their environment.

'Then wouldn't it be better to call those materials "wise", in the sense of an ancient oriental sage, rather than "intelligent", in the sense of shrewdness?'

Yes, the conditions of self-systematization that I have explained, emphasize functions which might better be called 'wise'. Furthermore, I would like to add as very important conditions, simplicity, an easy-to-control character, and the possibility of recycling, for the materials used by the next generation. An absolutely necessary condition is that they are gentle to the environment.

'Do you have concrete examples in mind?'

A paradoxical example of such a 'wise' material for the next generation might be traditional Japanese 'wood'. Wood as a structural material, adjusts to Japan's very humid environment by its capacity for self-diagnosis and self-recovery mechanisms without any help from microchips! I think it would be marvellous to find a new material which not only retains the merit of wood but also overcomes its weaknesses. Such a material, of which I dream, should be considered neither smart nor intelligent but 'wise', in the sense that it's full of wisdom.

'Now, could you explain what it means to be in the forefront of research?'

To be in the forefront is to propose an idea or hypothesis based on a certain fact, which will then lead to another theory. The fact alone will not lead on to further research without inputting new ideas which build on it.

'Could you give us an example of this?'

Take the humidity sensor I mentioned where we made use of a p/n contact. Ten years ago this was a major discovery. I was the first at that time to suggest that some theory could be made in this field. From the idea of giving intelligence to materials, we were able to progress to research in self-recovery, or the development of a similar p/n contact as a sensor for carbon monoxide; its modification led to a sensor which distinguishes between carbon monoxide and hydrogen. I suppose to lead in research is to extrapolate from one aspect of a discovery and develop it into a new field by looking beyond it. I tried not to carry on simply improving humidity sensors, but proposed instead the idea that they could become intelligent materials.

'It was then that you proposed the idea of intelligent materials and their development?'

Yes, I proposed that a new research field lay in that direction, that is the concept of intelligent materials and one example of this would be the p/n contact. If I had only proposed the p/n contact, it would have been too narrow a concept, so in order to make the concept more general, I suggested that it was possible to research intelligent materials from a different approach, although I didn't predict in detail what the other methods were.

'You mean you let other researchers follow your proposal step by step?'

Well, let's say I only hinted that certain fields lay ahead.

'I suppose in order to accept those new proposals, the research community needed to be fairly flexible. Did you get a lot of resistance at first?'

Yes, I was often told my ideas were ridiculous. When I mentioned that ceramics can have intelligence or can self-recover, they said ceramics were different from living organisms. Although I admitted this difference, I proved that ceramics do actually self-recover, and eventually my theory was accepted. On the other hand, people now say that they understand what I proposed, and I don't like this too much, because it means my theory has become too familiar.

'But that doesn't mean your research isn't at the forefront any more.'

Well, that's because it hasn't yet been put into practical form yet.

'So when that happens you have to move on to the next step.'

Yes, it's difficult always to stay ahead in the lead.

'In what circumstances do you usually come up with a new idea?'

It isn't something that just happens. I tend to try to understand certain premises or phenomena. It was after some trial and error that I came up with the first idea for the p/n contact. It flashed into my mind that it had connections with a living model. Before coming to this conclusion, I spent some years in discussion with people who had been studying the biological sensory apparatus, then I worked on it on my own.

'So, your thoughts need to ferment before you come up with a new idea?'

It's not a matter of getting an idea after repeated experiments; only after various trials could I come up with a hypothesis and then make experiments based on it. The results of these experiments lead to another hypothesis, so rather than pursuing the same thing for years, I was chasing what did not correspond with this hypothesis.

'So it's better to go in a lot of different directions?'

Not exactly. I think you need to have both ways in mind. That is, to have persistence and flexibility at the same time, because without the former, you might be distracted by 'common sense' arguments when challenged.

'So, if you think a certain hypothesis is true, do you have to stick to it?'

Not altogether, that would be conceited. You should pay attention to what others say, but it's your own decision whether to accept it or not.

'It seems necessary to have a certain kind of temperament. Tough and flexible for example.'

Yes, you need to find a balance between them. However, by saying this, I don't mean that these tendencies in any way cancel each other out. They have to exist together.

'Even when it means quarrelling with other scientists?'

Initially you're often seen as foolish but you need to stick to your ideas and not to withdraw.

'It's all over if you withdraw?'

Yes, that's the end of it.

'You have to keep on saying that what you're doing is the truth, and of course you need flexible research to support it.'

When you present a hypothesis and it is refuted by your own data, you have to be brave and admit that you were wrong. However, you must treat other people's data with care. There's no shame in overthrowing your hypothesis when you encounter other reliable data. You should be flexible if the data is good, because although it may seem to refute your hypothesis, it may in fact actually support it.

'It is important to get it right.'

You have to be persistent and thorough in your thinking; you can't skip stages when progressing towards a hypothesis.

'Do you see any special differences between the Japanese research environment and that of the West?'

It is often said that research in Japan is inclined to be applied, whereas pure research is done freely in the West, but this is not true. Professors in Japanese national universities are quite free but research in the US is rather like a game of survival. England has for a long time escaped from this tendency, though now it is in a similar situation to the US.

'Then the research environment in Japan is not especially different from that of the US?'

Of course there are many differences. The weak point here is that researchers have a tendency to flock around a subject that is already in fashion. For instance, research on superconductivity has now become very common in Japan, but the majority of researchers are mere followers. Another difference is that young scientists in first-class universities in Japan are virtually set free and left on their own while those in the US and in the UK are far more rigorously trained.

'What do you think about the Japanese government's policy in science and technology?'

This has also been misunderstood. It is true that government, especially the Ministry of Education, never supports university research, although they are maybe quite right not to because it hardly pays.

'Professor, you're the main advisor on MITI's next generation project. What is your policy for it?'
Only do what no one else in the world has done.

'Does MITI fund research much better than the Ministry of Education?'
Sure. They are also much freer in their support for research in industry.

'What is the attitude of Japanese industry towards your idea of intelligent materials?'
At first, there was no understanding, particularly for my proposal to make materials breakable for the purpose of recycling, or to make materials as simple as possible in order to promote technodemocracy. It is only recently that they have begun to move in these directions.

'I understand your research philosophy is that it should be done without constraints. Do you have similarly liberal attitudes towards women, or life in general?'
I try to see things from a free and unrestrained point of view, and I try to live accordingly. But when I meet someone from a completely different background, I often feel an appreciation for their different point of view. This then affects my research, and it enables me to judge from another viewpoint.

'I hear you are a great traveller around the world. Is this related to your philosophy of trying to look at things from different points of view?'
Maybe. Actually I could boast that the only place on Earth I've never set foot is the Arctic.

'Were you fond of travelling even when you were very young?'
Yes, but of course young children can't travel by themselves except in their own neighbourhood. So I invented for myself a sort of 'virtual' travelling.

'How was that?'
I made maps.

'Maps of actual places?'
Not necessarily. Very often I drew detailed maps of imaginary countries.

'What did they look like? Were they like real maps?'

Certainly. They had very detailed contours, cities, ports, canals, mountains, precisely determined roads and railways, and I even invented twenty-four hour timetables of trains running in my imagined countries!

'Extraordinary. Have you given up this pastime now you are grown up and able to travel by yourself?'

Not at all. I still do it in boring conferences! If I don't do the train timetables, it usually takes only half an hour to draw a map of a completely new imaginary country. I am sure I have invented a couple of hundred countries.

'So, that is a sort of "virtual" travelling.'

Yes, I imagine myself travelling around the country just by looking at the map I have drawn.

'Which countries do you prefer?'

I like India quite a lot.

'Why?'

Besides my interest in Buddhism, I am deeply attracted by ancient ruins, like Ajanta.[3] India also attracts me because it is diametrically opposite to Japan in some fundamental ways. I wouldn't agree that Japan and India are similarly 'Asian' countries. They are so different.

'What do you think is the major difference between Indians and ourselves?'

Mere vastness of nature makes people different. India has a sort of abundance which has nothing comparable in Japan. Human relationships there seem to be different and this is probably rooted in their attitude towards life itself.

'Are they for example more "cosmic" than us Japanese?'

Probably.

'I hear you are also very fond of Brazil.'

Yes. Again it is because Brazil is the opposite of Japan. You might say they live in much vaster space than us, meaning not only physical space but also mental space. It is probable that Indian and Brazilian people have completely different 'paradigms' of mind from us. Brazilians are also great artists. You see it even in their graffiti.

---

[3] A village in south India with caves containing Buddhist frescoes and sculptures.

'It is curious that you like these countries, because they are sort of "representative" of third-world countries. Their economic performance and the level of industrialization are well below that of Japan.'

I don't believe they are hopeless countries. On the contrary, they have great potential in natural and human resources. I don't see any reasons for them not to be great countries in the next century. Japan may have been just lucky or else astute in catching up with Western science and technology.

'It seems you see very little future in Japan!'

In some sense that is true. It saddens me to say that Japan is becoming a less and less interesting country. We have an incurable tendency to see trivialities as 'refinement'. We may some day be suffocated by our own preoccupation with detail! I hope that newly developing countries in Asia will not try to copy these aspects of Japanese culture, including technology.

'It is true that you are always trying to look at things from different view-points, just like looking at different countries. Do you like women to have completely different points of view too?'

Well, not only women. I like people with clear opinions, even if those contradict mine. It is fun to deal with someone who is a bit annoying if he means no harm.

'So for you, comfort is different from harmony in your relations with other people.'

Yes, it is. Like countries too. People who try to flatter are obvious and not at all effective. On the other hand, people who are straightforward are much more interesting. Maybe all these things have something in common—ideas for research, women, countries, values, and so on.

'If you were reincarnated, would you still want to be a scientist?'

That would be a scientific problem of reproducibility, but one can't go through the same experiences again, so it's impossible to tell if things would be done better. If we regard what cannot be reproduced as not science, human life will not be it, so ...

'Yes, it's true that one can't repeat one's life.'

In any case, a new condition would be created by trying to reproduce anything.

'As science develops, various things are explained but we can't predict our own futures. Do you think that's why we keep going, or would you prefer to understand everything?'

If I were asked, I would say carry on because I can't tell the future. There is always a new challenge, although not always leading towards a better life. It all depends on the individual what one lives for. But for me, I would say it is because one can't tell what's going to happen next. I mean I'm living because I'm looking forward to the next day thinking 'tomorrow will be a different world'.

'You mean it's interesting to look at alternatives among possible future worlds, is that it?'

Whatever it may be, it's better to know something new or find yourself in a situation with a different point of view. But I don't mind even if the next day is worse, since I've always been able to think 'now is the happiest moment of my life'.

YOSHIKAWA

EIGHT

THE SCIENCE OF ARTEFACTS

HIROYUKI YOSHIKAWA

TECHNOLOGY CAN be seen as a means of producing, utilizing, and maintaining artefacts. However, there has been little effort among technologists to consider technology from this unified point of view. Usually it has been conceived in terms of techniques and knowledge about specific artefacts, such as airplanes, cars, or computers. As a result, there has been a curious lack of concern about establishing a general theory or technology.

Professor Yoshikawa is the main exponent of this approach to technology. He graduated from the Department of Engineering at the University of Tokyo and first pioneered a 'general theory of design', which focuses mainly on the theory of designing advanced machines. This general approach enabled him to lead the development, in Japan, of two of today's key technologies, namely flexible manufacturing system (FMS) and robotics. However, his approach was so general that its scope has extended well beyond any specific mechanical systems. As a result, he now advocates that engineering science should be transformed into a more general science which he calls 'the science of artefacts'.

In this interview, he explains this philosophy in detail. Then, as he has been elected President of the University of Tokyo, the foremost 'device' which produces elites who control 'the system called Japan', he expresses his basic plan of how to reform this device to meet the challenge of today's world.

———

Although it may seem absurd, engineering has never been recognized as an authentic science.

'What do you think is the reason for that?'
First of all, no one can define what engineering is. There is, of course, marine engineering, mechanical engineering, civil engineering, and so on, which are separate sciences. But they are all very specialized sciences. You would hesitate before placing philosophy alongside shipbuilding. Philosophy can be defined as something general, like the search for the nature of human beings,

or of knowledge. But marine engineering is merely the design and manufacture of ships and other marine structures.

'Very few people have asked the question "What is engineering?", although many have asked "What is science?".'
Quite so. And I wonder why this should be the case.

'Because one of the characteristics of human beings is that they use tools and technology, and because engineering employs systematic knowledge of technology, the question of the nature of engineering may well be a part of the more general question of the nature of human beings.'
Yes, I agree. But I don't think there has been a serious attempt to answer this question. In general, the theory of science is still very young. Though we do have philosophy, which deals with human nature in general and more specifically logic as the science of thought, we don't have a theory of technology or engineering. Suppose a philosopher wrote a book on human nature. He would not spend more than half a chapter on technology.

'And so, what you are aiming at is a general theory of technology or engineering?'
Exactly. But I have not given it such an ambitious name. I have called it the general theory of design.

'Is this your invention?'
Yes, essentially.

'Is the reason you identify the general theory of design with a general theory of technology, because design is at the heart of technology?'
Yes. That is my fundamental hypothesis. The science of design has attracted much attention as a theory which can be systemized objectively. But so far no one has ever succeeded in building such a theory. In fact there are no 'departments of design' in any of our universities.

'Although there are such departments in art schools.'
Sure, but they are not dealing with science. Design in art schools is just a technique for creating artistic forms. Everyone understands its importance, and everyone knows that there's some element of design in any area of human activity. But there has been no systematic science of designing.

'Can I ask then how you came to be involved in the systematization of the science of design?'

When I was an associate professor at the University of Tokyo in the sixties, there was a massive dispute among the students. For a year, I was utterly prevented from doing any real research. It was a very tough time spiritually. I had a sequence of long discussions with the students. Because I was the youngest associate professor in the Department of Engineering, I was obliged to meet with the students from morning till night for four days in succession. At the beginning, there were a lot of professors there, but the older ones withdrew one by one, saying they were tired.

'You were the one who was left alone with them.'

Left alone for three days! I was forced to think about many different things not directly related to my subject such as the question, 'What is science?'. The essence of the debate was about the nature of science. The students complained about its one-way transmission from those in authority to students. In replying to their questions, I was forced to think about what I was doing myself. It was a harsh situation but I had a lot of time to think, especially because my subject, manufacturing engineering, was the target of attack in the sixties' environment.

'Like the way science was exploited by industry?'

I said things to the students like 'science is not yet mummified. It will benefit society in the long run'. But the students argued that university should not be a tool for industry to profit from.

'They accused you of being "bought" by industry?'

At the time I remembered the words of Professor Okoshi. What he said was that engineering science is not the industry itself. Its real subject is the intellectual creativity of people. Industry is the result of this fundamental human capacity. Thus industry is just one of the channels through which human creativity can benefit people. That was how I interpreted Professor Okoshi's thinking. I debated with left-wing students by resorting to this theory. What I noticed at the time was that intellectual creativity played a very important part in manufacturing. The subject of engineering science so far was concerned with manual or mechanical work. For example the whetting of stones or cutting of metal, how it is done, or how the process is automated. Or how to make devices create a desired surface. However, if you analyse the process you notice immediately, for instance, that when you manufacture a blade, the first thing you have to do is to imagine what sort of blade you want. This process is, in the vocabulary of engineering science, 'design'. So what is the nature of the act of design? I tried to find out if anyone had already written on the theory of design, but I was unsuccessful. That was in 1968.

'Wasn't it in this period that people began to be interested in artificial intelligence, etc.?'

Research on AI was budding at that time and people were trying to make a study of human thought itself, not from a philosophical point of view, but from that of information science. But there was no one in the AI field who was studying design. AI people were working on problem-solving, for example devising a program so that a computer can play chess or creating models of the process by which chimps reach for bananas. In recent terminology, the realization by computers of human intellectual activities as information processing. Rosenblatt proposed cognitive science and the modelling of the brain. But there was no one who tried to theorize about design in engineering. Of course there were many studies on design. But they were usually rather specialized, such as the design of ships, engines, other mechanisms. These were not related to each other. For example, architectural design is essentially a protocol for construction. I thought it was necessary to approach the deep-seated structure of human intelligence which controls specialized design activities. There must be some fundamental mechanism common to all human beings which underpins the act of design. I began the quest for a general theory of design in 1968. So my work was indebted to the student revolt of the sixties, in a sense, or at least coincided with it.

'So, in short, you tried to tackle the "essence" of what underlies the various design activities.'

I continued the project for several years, writing papers and campaigning in various academic communities. Around 1975, after many discussions with students and colleagues, I came to realize that the process of concept-formation in design can be interpreted by topology. This became a sort of central dogma in my theory of design. It was not clearly an idea of mine or anyone else: it came about spontaneously.

'Could you explain your topological view of design in plain terms?'

There are two topologies, though they are essentially similar: one is geometrical and the other is algebraic. We are usually taught at an elementary level stuff like 'a coffee cup and a doughnut have the same topological structure'. This is geometrical topology. This may be applied to database theory, but it is not a basis for my theory of design. The other topology is algebraic, which is much closer to my theory of design. In very simple terms, this is the concept of sets. Suppose there is a black circle and a white circle, you have two kinds of colour, black and white. Let's say, you then add a triangle. You then need a different classification to describe the group. In more general terms, if you introduce classifications to a set of several components, the set can be classified into many subsets adopting different types of classification and you will have a relationship or a framework of belonging where such-and-such an element belongs to such-and-such a classification. This framing process is called topos. It's not an unrealistic concept. The database in a computer is exactly this: data classified under several frameworks. It is sometimes said that this is in the form of a 'table'. This is the same as topos. For another example, the memory contained in our brains may be seen as equivalent to a table so it could be said that our memory has topos.

'We naively believe that concepts like "black/white" and "round/square" exist for what they are. But what you are saying is that they provide an abstract framework into which crude data are recorded and classified.'

Exactly, even in the natural sciences. Take, for example, the system for classifying living things. We may think a classifiable entity, such as a crocodile, is a natural species, but this is essentially hypothetical. There are many sorts of possible classifications dependent upon fundamental hypotheses.

'Classification, then, is always a creative view of the world.'

To imagine intelligence as an accumulation of memory, this would be a rather one-dimensional dictionary definition. For instance, let's imagine there were several sorts of meat in front of a primitive man or woman. Let's say there are three: one is fresh, one is rotten, and the last is completely dry. If they were presented to a dog, the situation would be quite simple. It just takes the fresh meat and forgets about the rest. However, for human beings it's different. They will examine each option. After experimenting, they will start 'classifying' them. From a viewpoint of edibility, the fresh meat stands out from the others. However, the human soon notices that the fresh meat will also rot eventually. Thus, from the viewpoint of durability, fresh meat and rotten meat are similar, while dried meat stands apart. It can't be eaten, but it won't change in time. So a primitive person begins to wonder if there is a kind of meat which is both edible and durable even though he doesn't have it in front of him.

'A sausage for example!'

Yes. Something new emerges in a situation where two different viewpoints intersect.

'Do you see the birth of planning or design in this sort of situation?'

Yes. Design is based on recognition. But it is an act which goes beyond recognition. The fundamental principle of creativity in design is to see an existing framework as something to be acted upon, and then to propose new concepts.

'This is an interesting view. Could you explain it in more detail by giving us concrete examples? Could your act of design be imagined as for example building an optimal ship by calculating the dynamics of the waves, requirement for load, etc.?'

No. My view is more abstract or structural than that. First, the act of design has three different levels. If the requirement is to import rice from the USA, the first level would be to take account of social requirements such as what is the most efficient way of bringing rice into Japan from California. There are many degrees of freedom here. When and where do you get rice? Where do you polish rice? You have problems with social institutions like customs. And would you carry the rice by ship or plane? This is the first level of design. The second level would be planning the shape of the ship. And finally, after you decided the shape, you come to the last level of design such as what sort of steel do you use for the ship and how much?

'The last is the level which we usually call design.'

Yes, but what I call engineering design has three levels or stages.

'The first stage looks the most interesting.'

And the second is essentially a matter of computation. You have to resort to sciences such as hydrodynamics, materials science, and make use of high performance computers like parallel machines. Of course current computers have many limitations; for example, the problem of turbulent flow is insoluble even by supercomputers. But the structure of the problem is clear for the second and third level of design. The first level of design, however, is not for computation.

'You are dealing with society itself.'

Yes, indeed. Moreover, the parameters range from the technical to the social. Also, one has only a vague idea what sort of strategy people have in mind when they are dealing with these problems. I attempted to approach this problem using concepts of topology.

'You have applied your problem to concrete problems, especially robot design and precision engineering?'

Let me take one small example. There was a product sponsored by MITI called FMS (flexible manufacturing system). In general, automization is what makes mass production possible. However, in normal mass production, we only have standardized products. But people really want their products tailor-made. FMS is a project which creates a completely automated, diversifiable production system. In leading this project, I was engaged in the conceptual design problem of defining the system. Roughly speaking, we had one category of problems, that is the social and labour structure including job training. On the other hand there was another category of purely technological problems such as materials and energy. Hitherto, these two categories had been treated separately. Job training and the energy environment were kept apart. But from the point of view of topology we had to discuss these problems together to define the topological structure. Suppose you have twenty factors, then only after those problems have been related to each other, or their topos has been defined can you determine the final design of the production machine. In fact, FMS has become quite famous around the world and many people have come to see the system. The actual product is a system consisting of manufacturing machines which have been put together. They are completely automated, though highly flexible. This FMS project helps Japanese machine industries to improve their competence in design.

'After automation, there is only one step: to robotics.'

Yes, robotics research in my lab began with the question of what kind of robot was needed but which did not exist yet. We looked at modern society as a whole and asked what essential robots were lacking. This is the way I carry out any project, a top-down approach. We talked both inside and outside the lab, posing this problem, 'what is the robot of the future?' We have to follow the trends that society is moving towards, after which we have the problem of labour, and then of course there are serious psychological problems. We looked into these factors. After discussing them from 1978 for several years and structuring the problems, we arrived at the new concept of 'maintenance robots'. Until then, there had only been manufacturing robots which had reached saturation point, but there was nothing which could do maintenance work.

'Maintenance robots diagnose and repair the defects, the diseases of the system.'

Exactly. They are doctor robots. The main premise is the social situation. We are increasingly surrounded by artificial environments. But the essential feature of any artefact is deterioration. Machines are best at the point when they are created; any subsequent changes take the form of deterioration. We thought that this problem would be the point of maximum contradiction in future society. Out of these arguments, together with the design theory, we proposed that maintenance robots would become necessary in the fight against the problem of deterioration in artificial systems.

'There must have been many interesting applications.'

One example was that we proposed maintenance robots which would dwell inside nuclear reactors, repairing defects immediately on site.

'The concept of maintenance, I think, concerns the safety of technology. Is that true?'

Yes it is. And I should add that the concept of maintenance is also related to biological systems. I was interested in the difference between the idea of safety in biological and artificial systems. As you know, living things have an amazing capacity for self-repair. Industrial plants don't at all have the same capacity. You might wonder if artificial systems have some fundamental flaw. I am not very well informed about biological systems, but I understand that they have very fine mechanisms to maintain themselves. On the contrary, artificial systems don't put any effort into self-repair. After you build a machine, you leave it to itself and don't do anything more until it breaks down. Then you might have an accident. The reason we have a high incidence of accidents is that the artificial systems of today don't have much in the way of built-in maintenance systems.

'There may be economic reasons for this problem.'

Before I launched the project, I looked into the statistics to see how much of the gross national product is laid out on maintenance, for example, the fixing of airplanes and cars, and the maintenance of railways or steel furnaces. We found that the total cost was five per cent of the gross national product. We were convinced that this cost would increase. Or more to the point, we have to increase it to prevent accidents.

'Is this also related to environmental issues?'

In principle, if we are to develop any new technologies, we have to evaluate how much maintenance they need and take the cost of that into account. As you may have noticed, this has not been taken into consideration in technological assessments. I would predict that the more technological inventions we have, the heavier the burden of their maintenance becomes. As a result, this burden will crush industry because the pressure of maintaining these artificial systems will overwhelm the economy. This is the fundamental defect both in methodology and in the design of the current technological systems.

'You have to design a system which can be maintained.'

Yes. The act of design, including designing for maintenance, will gain much more weight in the future. I think a great clue comes from genetics. I would think there has been no estimate of how much cost biological systems have to bear just to maintain themselves. So, it is out of the question for engineering scientists to use terms like 'biomimetic' easily. The reason I am attracted to genetics is that the genes are a sort of design concept which is very simple and is not a collection of subsets. We have to start from design, for example, of very tiny creatures with intricate genes.

'Yes, I was thinking something along those lines. There are various different types of animals but the fundamental design concept is very similar. It's the design of life, isn't it?'

When I presented the problem of maintenance earlier, I said that we have to learn from the physiological functions of the body and not the information functions of the brain. Engineers in general tend to think that we have to learn from the brain and then implant AI into artificial systems. But there are many different aspects to biofunctions. For example, the natural repair of wounds is nothing to do with consciousness. Creatures do a lot of things unconsciously. In designing maintenance, we need to learn about these biofunctions.

'It's much more interesting to try to see what we can do without reference to the brain. That's what you are suggesting! So today's environmental problems point up the brain-body conflict you describe. It is not the brain but the body which is more sensitive to the environment. Air pollution affects the body much more than the brain.'

Modern, or Western, science and technology have followed leads which inclined more to the study of the brain. The science of artificial systems operates under maximally optimum conditions which are calculated by the human brain. This is imposed on nature because artefacts become a part of nature the moment they are created. The logic of nature is not exhausted by human intelligence because there are an overwhelming number of unknown factors which are proper to nature. To control the process totally, we first need to have design. This is done by the brain. The next thing we need to have is maintenance. We have to design a system in which maintenance is performed unconsciously. Otherwise, artificial systems would not have the same stability as biological systems.

'There are a lot of essential jobs for a system that seem wasteful or foolish to the brain.'

In that sense, maintenance must have a totally different logic, or paradigm, from previous technological systems such as the creation of aeroplanes and cars.

'Can you expand your thoughts on this paradigm change?'

Let me begin by affirming that modern technologies have been founded on the basis of modern science initiated by Newton. Taking for example the famous anecdote of 'Newton's apple', we have been told that Newton discovered the law of gravity when he saw an apple fall from a tree in his garden. True or false, the real meaning of this story is not very clear. The moral would seem to be that he discovered an abstract law after contemplating deeply an actual phenomenon taking place before his eyes. It may also mean that Newton was such a genius that he got inspiration from an event as trivial as the falling of an apple.

'Do you have a different interpretation?'

I doubt that the anecdote is factual. But, putting that aside, I imagine that Newton used it intentionally to make his point. What he wanted to say in employing the story, I think, was that two such remote phenomena as the falling of an apple and the movement of the stars can both be described by his very simple law of gravity. In other words, he showed the consistency of his idea by extending it from apples to stars. The central idea behind his *Principia* is this consistency. That's the moral of the anecdote. The same law presides over two very different bodies.

'He was not so much interested in explaining why an apple falls, as that apples and stars are the same under his law.'

Yes. And my thesis is that the fundamental assumption behind the science of artificial systems is that Newtonian consistency does not hold. The difference between the science of artificial systems and Newtonian natural science is the difference between finite and infinite systems. In technology or engineering, we are working on the tiny area called the earth and experimenting with things which can be intentionally controlled. So our experimentation is finite. All artefacts are perishable. That's the essential condition from which we suffer. We are living in a finite world surrounded by infinity. Nevertheless, we have to come up with effective rules. So the Newtonian paradigm, the law presiding over infinity, is not valid. We must ask Sir Isaac to retire from the scene and then create a theory which is effective in the finite world of artefacts. This is the science of artificial systems. Do you agree that they have different paradigms?

'You are saying that with artificial systems, we cannot expect that what is true in the infinite world, or in nature, is valid.'

Modern natural science presupposes infinity. For example, what you see in special circumstances in accelerators can occur in exactly the same way in black-holes far out in the universe. That is in Newton's paradigm. What we need is a logic for closed or finite space in the world. So I suggest that engineering science is the science of the finite. You can in principle experiment endlessly in natural science. But suppose you experiment on an industrial plant so that it explodes! The number of plants we have is finite. You have very few, or no sample at all, to experiment on. Nevertheless you have to discover under what conditions it will explode. You can't experiment on a nuclear reactor by blowing it up. It would be the first and last one. This kind of situation is outside the realm of natural science.

'Everyone knows that we get old and die. No one expects a man of seventy to behave as though he was thirty. But we are demanding that the nuclear reactor doesn't age or die.'

Yes, exactly. *That's the illusion of infinity.* Though it is a natural science, biology may have more of a flavour of what is finite, but molecular biology, I think, has no interest in it at all.

'No one has much awareness of the perishability of artefacts. Everyone demands of machines that they don't break or age.'

That's in the wrong spirit of science. It's an illusion that it is exceptional to have a life-span in the infinite universe. All creatures will eventually die. There's no exception to this. Of course all artefacts too will pass away without exception. But people hear a creak and say it shouldn't exist. Why? They suppose that what is made will last as long as the manufacturers want. This is wrong. As Democritus said, everything with form will perish. But natural science neglects the life-span of its elements. Of course, the atom has a life-span. But scientists don't care. It might be called an 'infinity cult'. Modern thought is dominated by the belief in infinity as an ideal concept. Finitude is seen as some sort of exception, or anomaly. But it's actually the other way round! The fundamental presumption of engineering, the science of artificial systems, is not the same as that of modern or Western natural science.

'I think this is a very important issue in establishing the fundamental difference between science and engineering. Are there other ways in which engineering is different from science?'

Let me take one example. The most important function of science is the discovery of new events, matter, or natural laws, that is, in short, novelty. Scientists are people who break into the unknown. So analogous to this, the conventional idea of engineering is the uncritical supposition that the function of technology is to invent 'something new'.

'An American way of thinking?'

Possibly. However, technology or artificial systems have nothing to do with novelty. What we value in technology is the overall balance of the system, simplicity, friendliness to human beings and the environment, and so on. The impulse of technologists should be different from that of scientists who are eternal novelty-seekers.

'I think all those issues might be seen from the viewpoint of "modernity" or "modern Western civilization". Could you talk about modernity in science and technology from the viewpoint of Japan?'

I think, in the case of technology, that Japan has been more Westernized than the West itself. There are a lot of issues here. But, from the viewpoint of social design, one of the most fundamental inventions of the West is specialization, or the mass-production of specialists. This was the basic mechanism which enabled the tremendous advance of science and technology both in the West and Japan.

'In ancient Asian religions, such as Buddhism and Taoism, the most important thing was that individuals reach the whole truth of the universe.'

The West ousted such obscure philosophies and invented a method of accumulating knowledge and technology through the cooperation of a vast number of '... ists', such as physicists, chemists, and economists. I would point out that this theory of over-specialization has become practical wisdom. In other words, the 'problem' has been defined clearly.

'Have you any solutions to this?'

Of course I don't have a magic solution. Though I have a number of ideas. Let me explain one thing. The fundamental driving force behind the advance of modern science and technology has been the liberation of the desires of individuals. Poverty has been defined as the difference between the reality and desire. The experiment of socialism of the Russo-Chinese type was an attempt to surpass modernity by suppressing these desires. But it ended in complete failure. Thus what we are facing now is the domination of the market-place as a field of naked desire. I am wondering if capitalism has the power to resolve the contradiction between these infinite desires and finite resources. I would doubt it in as far as we measure richness and poverty by the quantity of the products we possess.

'Do you have an alternative proposal?'

I would propose that we measure richness not in terms of things but of service, and not by possession of things but by the provisional possession of service.

'"Virtual possession"?'

I would say that the current system of mass-production and mass-consumption is bound to cease eventually. People must learn to not own a lot of consumer items, but to explore the services available in a tentative and functional way.

'How would you design such a system in practical terms?'

I think it is only possible after the transformation of the economy and technology. But the range of possibilities may be very narrow. I can't propose any concrete plans right now. But I am sure the present system won't work very long.

'I would like to change the subject now and ask you a little more about your thoughts on higher education in Japan, because you were elected President of Tokyo University this year.'

Our universities, especially the University of Tokyo, were instituted by the Meiji Government as mechanisms for importing Western science and technology into Japan. They have been very efficient, even more efficient than universities in the West as machines for producing the '... ists', the specialists we have just discussed. But these machines are now out of date and need to be remodelled.

'Do you intend to scrap Todai,[1] then?'

Todai has been criticized as symbolic of the notorious Japanese 'hell' of entrance exams, the suppression of creativity, and the fostering of intellectual authoritarianism. In this sense, I agree with the scrapping of Todai.

'Please don't forget your words will be translated back into Japanese!'

But not right away. And I suspect Todai is also a symbol of the exceptional eagerness to learn of Japanese people. Todai has been a centre for this beneficial potential. I don't want to suffocate the Japanese desire to learn.

'Then, how would you go about it?'

We have to explore the influence Todai has on primary and secondary education. After that, we have to transform Todai gradually.

'In what direction?'

I don't agree with our present system of mass-producing sinister '... ists'. The university should be a place for the accumulation of 'wisdom' in society. Its aim should be to devise solutions to the problems which our modern global society presents us with. Until recently, Japanese universities have been concerned with importing and monopolizing Western knowledge. But I am determined to transform the universities into something more open and essential. I don't want universities to be machines for absorbing clever people and manipulating them to become '... ists'. The university should be a means of generating wisdom.

'How long do you think it would take for this to be realized?'

About ten years. We have to do it by the beginning of the next century.

---

[1] Tokyo University.

KARUBE

NINE

BIOELECTRONICS AND THE ENGINEERING MIND

ISAO KARUBE

THE ELECTRONICS revolution started in the field of communication and then proceeded on to computation and control. Its progress has been so rapid and sweeping that there are hardly any human activities which are exempt from its influence. There is, however, a vital field to which electronics has only elementary access at present. This is the human body itself. Would it become possible in the future, for example, for the brain to control directly artificial legs using an electronic interface? This research field is sometimes called 'bioelectronics'.

Professor Karube of The Center for Advanced Science and Technology at the University of Tokyo is one of the world leaders in this new field. He is famous for having made working 'biosensors', the electronic sensors which utilize biological elements. He is now engaged in a number of challenging projects including the development of biochips, a new class of semiconductor devices which utilize biological molecules.

In this interview, he talks freely about his research, his philosophy, and his dreams for the future.

―――――

'Your first biosensor was an enzyme sensor, which was not actually very "biological", because enzymes are protein molecules. But then came the real biosensor which used living microorganisms.'

After developing an enzyme sensor, I invented the first biosensor using microorganisms in 1974. This was our original work and we were the first in the world to put a sensor using living organisms into practice.

'So this was one of the earliest attempts to bring living organisms into the world of electronics?'

I think so. Our first target was to build a BOD[1] sensor so that we could make use of microorganisms. I have never been able to get the problem of

―――

[1] Biological oxygen demand.

the environment out of my mind, because as you know, BOD is an important barometer for measuring how polluted rivers and factory effluents are. Before my sensor came into practice, it took five days to measure BOD. I thought if we went on like that, we would never be able to monitor effluents, and therefore I should work on a BOD sensor, in which I was successful. Today, it has been standardized by MITI.

'How did the microorganism sensor come into being after you had developed the BOD sensor?'

Alcohol sensors and acetic acid sensors already existed, and also a sensor for glutamic acid measurement.

'What are these actually used for?'

For monitoring fermentation processes. For instance, Ajinomoto, a pharmaceutical company, and I invented a sensor using *E. coli*, and now it is used for process measurement.

'What sort of microorganisms are involved?'

We use various kinds, especially yeast. *Trichosporon* is often used also. For the BOD sensor and for a sensor used to measure acetic acid and alcohol, we used *Trichosporon*. It is very durable and can be used for up to a year if it is treated delicately. It is very stable as though it wasn't biological. It lives inside the sensor and reproduces so that it can be used for a long period.

'I suppose your range of interests have extended beyond enzyme and microorganism sensors?'

My current research is on how engineering can tackle the information processing function of living organisms of which biosensors are just one example. As a result, I am still working on various biosensors and biochips. I have invented a chip consisting of a pigment protein called bacteriorhodopsin which exists inside the cell membrane of halobacteria, combined with an FET. When light strikes bacteriorhodopsin, a gate opens and hydrogen ions flow in. This affects the transistor and creates a weak electric current, so this chip can be said to resemble a switch with an on–off function. If we could realize a function such as that of bacteriorhodopsin using a single protein molecule, we would have an extremely small semiconductor device which had never been thought of previously.

'This is the direction where devices are made out of biological molecules. What other directions are there?'

I am looking into gallium arsenate related research. The aim is to make the sensing elements themselves into molecular devices, I mean, bringing them down to ultramicro levels by employing processing technology for microdevices. By making these sorts of ultramicro devices, I am thinking about their application for information processing, and wonder if they could be adapted to various forms of so-called biocommunications or bioarchitectures. This is my main research.

'There seem to be many facets to your research, but the main point is that you are trying to look at the biological world from an engineering point of view, and experimenting with ways of bringing them together.'

I was taught to have an engineering mind while I was at the Tokyo Institute of Technology. The basic ideas of practical engineering were impressed on my mind, that is, that the engineering department would be of no use if it didn't take on work that was useful to society. Also, I was taught not to be a follower, that is not to do the same research that others were already doing, but rather to be original. This ideal was emphasized. For example, if I came up with an idea the first question was whether anyone else had already thought of it. There was a consistent ideal of doing work that had real use.

'I think, however, that the concept of the enzyme sensor had already been proposed in the USA?'

Yes, research on enzyme electrodes started in the US, but the reason I am recognized worldwide is probably because I extended the concept of the biosensor. Through the development of highly original sensors such as the microorganism sensor, the DNA sensor, the freshness sensor, the multi-functional sensor, the micro-biosensor, and the ultra-microscopic amperometric sensor, I have extended the field of the biosensor. We are now the largest group doing biosensor research in the world. But when we began our research, there were only two or three people in the world who were doing this kind of work.

'Where were the other researchers?'

They were from the US. There weren't any in Europe.

'What was their work like?'

They were doing research into sensors using enzymes, mainly for medical use. Among them, Dr Gilbert was researching comparatively widely into electrode sensors using enzymes. I consider him as the person who made widely available the use of the enzyme electrode. However, he started his own company and retired from front-line research. Drs Updike, Hicks, and Clarke have also retired from the front line, and we are now the only group which has devoted itself consistently from the beginning to research into biosensors.

'Of late, there seem to be researchers in Europe too.'

The United Kingdom has made progress, and Germany and France are following hard on their heels.

'What are the characteristics of research in the UK?'

They are mainly working on medical diagnosis sensors. They have succeeded in making practicable disposable glucose sensors. However, they don't have enough tradition yet in the field, so they may not be good at what we call high-technology sensors.

'Is the field growing in Japan?'

Japan is first in the field, taking into account the number of participating companies, the number of patents, and the number of published papers.

'Is this a field in which companies might be interested?'

Universities are certainly doing well. Interest among companies is also quite high, and there are nearly ten which specialize in biosensors, because they are relatively quick to be put to practical use. While pharmaceutical drugs are said to take more than ten years before they become commercial products, some biosensors could be in practical use within two to three years; this may be the factor that encourages companies.

'I see, and another factor I presume is that they don't go directly inside the human body.'

Yes, I suppose that's why there has been such rapid growth. There are consistently more than ten people from different companies working as visiting researchers in my laboratory.

'Could you give us an interesting example?'

Well, we have a toilet maker who is researching into disposable sensors, calling it an 'intelligent toilet'! The idea is that an integrated sensor is built and placed in the toilet, so that it can detect the health of a user immediately after it has been used, for instance the amount of glucose within the urine.

'The need for daily checking has increased because of what we have found out about geriatric diseases.'

For instance, it may be difficult to detect glucosuria after the toilet has been flushed. But if a sensor measures the amount of glucose in the urine, it is then possible to detect diabetes at an early stage. Furthermore, a sensor is being developed that can examine various parameters such as the amount of protein in the urine, or occult blood in the faeces which is the first symptom of colon cancer.

'Could I ask what is your general view of the development of biotechnology?'

Imagine biotechnology as a Cartesian plane. Take the development of biotechnology from the biological point of view as the vertical axis; and from the technological point of view as the horizontal. The vertical axis starts out from the rearrangement of the DNA of prokaryotes, and eventually progresses to eukaryotes. The progress of 'biotechnology as biology', including the genetic engineering of higher animals in a wider sense from chromosome engineering to embryological engineering, occurs on this axis. On the horizontal axis you will find research into developing artificial devices out of living systems. One of the ultimate goals of this direction, as I said before is the molecular biochip.

'Where are your biosensors located?'

They probably lie half way in between living and artificial systems on the horizontal axis; they can be seen as the interface between living systems and electronic systems. At the moment, research is more active on the vertical axis, I mean even now we are at the point of imitating biological functions such as producing hormones by genetic engineering and introducing foreign genes into host genes. But from now on, when engineers seriously participate in this field, I think the horizontal axis, from living to artificial systems, will become the main stream.

'Not imitating biological functions, but making use of them in artificial systems.'

Yes. For example, when we operate neurocomputers, it is not by using the actual network of neurons itself but by using a software that uses the same logic as neural functions. Neural networks may be realized by parallel processing using neurochips. We call a device that is good at parallel processing, which is appropriate for such neural networks, a neural chip. Though we start from the imitation of biological functions, we are going into the world of artefacts, and departing from the biological world.

'The biofunctions are sort of "extracted" or borrowed to construct devices which will make up the system we desire.'

Speaking loosely, a 'function' such as this would not exist without 'analog' creatures like us. So we need to make a device which doesn't use the nerve cell itself, but is made from it or, if necessary, from some of its molecules, thus making an electronic device that can run software. This shows that we are apparently moving from living systems towards artificial systems along the horizontal axis. To extract information processing mechanisms and programs from living organisms and then realize these functions artificially would be the work of engineers. The progress of biotechnology as biology and of biotechnology as technology seem to have completely different vectors.

'In which direction do you want to proceed?'

Naturally, as I am trained through-and-through as an engineer, as I have already said, I would carry on research along the horizontal axis. My main aim is to apply the superior functions of living systems to engineering. This is my basic research position. Specifically, I like to work at the interface between electronics and biology. I described earlier, concerning this interface, how to make the brain communicate with computers, or how to measure chemical information in living systems with a sensor or a sensing device, converting it into electronic signals.

'You mean the information carried by transmitters?'

Yes, neural transmitters are a case in point. For instance, a biochemical transmitter is released when an impulse arrives at the end of a nerve, but if we could sense this with a molecular sensor we could convert it into electronic signals, that is if we could receive electric signals artificially by biosensors or a molecular device. If we could establish the interface between biological systems and electronic systems using molecular devices, there would be the possibility, for example, of moving artificial legs with the nerve impulses. That would be the exact interface between electronic and biological systems. This is what I am most interested in.

'Considering the way society develops, computers will be supporting human beings. This will be in ways like the one you just mentioned where biofunctions are imitated, using neural networks, software, and neurochips. On the other hand, mechanisms can be developed in which the living system is the inferior part. Then changing the living system functions back to biosystems by the method of interface you suggested would develop something even better that could not have been realized in nature?'

That is certainly a possibility, since living systems are not always superior, for there are a lot of things they are not good at. Taking our five senses for example, our hearing is quite limited. We can only hear what is within the acoustic range. There is a story of a horse being 'shot' with a supersonic gun during a horse-race in England; even if this was not a true story, each animal does hear in a different frequency band. We humans recognize sound through a so-called sonic sensor in the ear, but the range is limited.

'Can electronics overcome these limitations?'

Electronic devices have a far wider dynamic range, so by using electronic devices to look at the wave form we are able to pinpoint an approaching sound, and tell which direction it comes from, even though we cannot hear it. In other words, this is a use of technology to improve sensory capability; we can do what is beyond our natural abilities thanks to electronics. There is another side to this, though. For instance, let's say a group of people are talking near to you. It's noisy and you can't clearly hear who is talking, but if you listen carefully, supposing that one of them is saying bad things about you, then one voice will stand out from the others.

'The famous "party effect".'

Yes. The brain picks up one particular sound by adapting the frequency. This cannot be done even with a high efficiency microphone. Thus, both living systems and machines have merits and demerits, so I believe feedback between the two enables the progress of technology. Of course I don't mean that a machine cannot do this at all, for when it recognizes the characteristic of one sound from various noises, it can pick up that specific frequency.

'This, for example, is what a voice input computer does.'

It is possible to preserve one specific sound and filter out the rest, but in order to do this, there are problems which can only be solved by supercomputers! In any case, technology has actually begun to support the weaknesses of living systems.

'So we will be able to sense things which have never been sensed before.'

For example, Kiko-jutsu[2] is now in fashion. Even I could move a static piece of paper with my force, like this! This energy might possibly be measured by a sensor, perhaps a quantum wave sensor that works on a completely different theoretical basis. Now that people's attention is turning towards the inner world, in the developed countries where materialism has reached saturation point, the future of electronics depends on the problem of what sort of approach to take towards the brain, the neurons, and the mind.

---

[2] An Asian discipline which develops inner energy called Ki.

'Is there anything that you are already studying from the viewpoint of the interface between the brain and electronics?'

The problems of the brain, especially the problem of dementia, are actually being researched. An ultramicro-biosensor has been developed. Since this sensor is microscopic, it can be put inside the cell without killing it, because it is covered with a biocompatible membrane. Using this ultramicro-sensor, we are trying to work out the substance flow inside the brain.

'And has this been successful?'

The basics have been solved, so I suppose it is now a question of time. The sensor we invented is two microns, but we are trying to bring this down to 100 angstroms. Since cells are about twenty to thirty microns in size, the current two micron sensor can detect information such as the release of transmitters, or the dynamic increase of a substance when the brain is memorizing. This is the first goal in creating an interface between the biosystem and the electronic system. First, we decided to trace the flow of glutamic acid in the brain. In order to do this, we put a bio-adaptable material on to the surface of the sensor, but because the sensor is so small and the amount of enzyme that can be fixed in it is limited, we are trying to achieve a technological improvement.

'So you will be using an enzyme sensor?'

For the glutamic acid, I am using an enzyme. Acetylcholine could also be sensed with an enzyme. Immunosensors are used for the rest. Recently, a researcher from the Tokyo Metropolitan Research Institute of Ageing came to us and proposed that he would like to apply them to research into Alzheimer's disease. In this disease amyloid precursor proteins accumulate inside the brain. At an early stage, amyloid seems to be still water-soluble. The researcher mentioned that he has an excellent technique for operating on rat brains, and then putting sensors in to them. By putting a sensor into a cell, or between two cells, I mean at the synapse of the neurons, we are trying to find out what sort of information flow occurs when an impulse comes, or while the brain is memorizing. I think it will be possible to do the same with the substance that controls sleeping. This research on ultramicro-biosensors is our major concern.

'Could you expand on this?'

A carbon fibre is used in the sensor we are currently developing. The width of one sensor is two microns and we are now in the process of investigating the characteristics of the sensor. Furthermore, as I have said, I am trying to make a submicron sensor by micro-fabrication technology. For instance, a micron line is created with gold and made into an electrode. Since a counterelectrode is needed, you then create another submicron line elsewhere. After putting the electrodes in an enzyme-polymer solution, the enzyme will attach itself to the electrodes. The enzyme reaction will make it possible to measure how much substance is between the electrodes. Nothing would be better than if we could create another FET. If we establish electrical potential with FET, it can measure substance using enzyme reaction.

'In order to measure minute concentrations of a substance, you really need to measure a tiny electric current?'

When we measure an electric current, the electric intensity is probably a picoampere or less. So the circuit design is the hard part. Today, a nanoampere current might be caught, but in order to catch a picoampere current, we only have to extract a signal with an electric current. If you try to do this with a potentiometer, nerve impulses often get in the way in the form of noise. A circuit, completely controlled by a computer, has been made, but the biggest problem is what substances to build the submicron sensor on! It would be best if we could attach submicron gold on to a silicon of one millimetre thickness.

'You mean that submicron gold becomes the electrode?'

Yes. We put the electrode on a support and then into a cell or synapse. This sensor is not just for the brain, but could also be used in blood tests and thus enable us to do a test that would be felt hardly more than a mosquito bite. Since there are thousands of people who need to have blood samples taken every day, this would relieve them from pain! I call it a 'mosquito sensor'. In any case, with this ultramicro-biosensor, the physiology of the brain will be revealed in great depth.

'So far we have been unable to look at the brain from outside.'

At the moment, what can be done is either to measure the impulse or investigate what is extracted from the brain. The most advanced technique is to detect positrons which are emitted from inside the brain. But by using this ultramicro-biosensor, and putting various sensors such as those for glutamic acid, glucose, or dopamine into the brain, we can take direct measurements from various sites and watch how the flow of these molecules changes, or what sort of molecules enter into which area of the brain as one exercises, thinks, or dreams. Through this, I expect that mechanisms of information processing inside the brain may be uncovered. I think brain function expresses itself as the flow of molecules, that is as chemical information; from this point of view, the brain cannot be understood unless the dynamic flow of molecules is traced in detail.

'Not to mention the highest functions of the brain.'

This sensing would be an interface at the molecular level. Unless an interface between the biosystem and the electronic system is devised at the molecular level, I don't think the function of the brain will be understood.

'What will happen when such a superior interface is realized?'

My wildest dream is to create a form of biocommunication using sensors that measure energy at a different dimension, such as quantum wave sensors. Of course, if it becomes possible to put sensors easily into the brain, I suppose psychological activities such as memory, thought, and emotion will be sensed electronically.

'The transmitter profiling a person?'

If it becomes possible to sense the flow of transmitters in what is regarded as a modern disease such as manic-depressive psychosis or stress, we could view these diseases using objective data. The day might come when we could decide whether or not to work by observing the dynamic flow of transmitters in the brain! By integrating sensors for different transmitters, a 'communication cap' could be made which would enable us to communicate with others without using language!

'You might be able to read other people's minds?'

Yes, brains would be able to communicate directly without using language.

'This would be a world of pure communication with no difference between principle and real intention.'

Even further into this SciFi world, not only sensors, but devices or interfaces that could replace brain function and be directly connected to the brain may be possible.

'What would you like to do now in your research into other interfaces between the biological and electronic systems?'

One thing that I would like to do is achieve a chemical interface with plants. The way plants process information without nerves or brains is quite fascinating to me.

'Plant information is a field that scientists have recently become interested in. For instance, it has the characteristic of totipotency, that is, no matter which cell is taken out, it creates an individual, and the way it conveys information differs from that of animals.'

What is more, the floating substance is of picogram scale, from which buds and flowers grow.

'And it is said that plants quite far apart can exchange information.'

They seem to communicate with the atmosphere around them by means of molecular sensing. Recent research has only concentrated on neurons, but it seems to me that plants may be better at chemical communication than us!

'Probably because they have much less to say?'

The basic logic of the information processing of living systems lies in chemical communication, on which plants rely! Animals have chemical communication too but it is too complicated. We animals have to study plants, and come up with a theory about the language they are speaking. From an evolutionary point of view, of course, animals originally descended from the same ancestors as plants, so I suppose the general grammar of a common language could be discovered up to a certain point.

'The biocommunication with the brain that you described is interesting, but biocommunication with plants seems to have a greater reality.'

Don't you think it would be marvellous if we could receive the soundless signals that plants are emitting, such as 'I want phosphorous', or 'I want nitrogen'?

'Or, "I don't want to be cut down"?'

Well, perhaps. I also think it would be possible to convey our will to plants by creating a synapse with biochips, like a modern version of 'The blooming old man'.[3] One could climb up a cherry tree that is not in bloom and give it an order to bloom, so that it detects the information and flowers grow! Seriously, though, biocommunication with plants based on genetic engineering and bioelectronics may become the key technology for solving food and environmental problems in the next century.

'You said that Japan is the leader in biosensors research. As a leading Japanese technologist, do you have any message for the world?'

The tendency to think of Japanese technology as being without originality is completely outdated. Today, the technology of Japan, the US, and Europe is almost at the same level, and the problems they are facing have many things in common. To put it very simply, I think the essence of these common problems is how to transform the materialistic civilization that is characteristic of the twentieth century to a spiritual civilization in the twenty-first century. Recent research, for example, into the electronic interface with the brain or

---

[3] An old Japanese story about an old man who makes flowers come out of a dead tree.

chemical communication with plants, is a way that we technologists, worldwide, have of changing our materialistic civilization.

It has been said that it is difficult for the West, with its Christian values, to understand the different mind of Japan and the rest of Asia. However, it is possible that two different currents could merge under the influence of the development of a universal technology. The Japanese have been criticized for taking over research developed elsewhere, but in the research to establish the cornerstones of a spiritual civilization in the twenty-first century, I would like to see Japan as sponsor and leader, and to let the world take advantage of this.

BEPPU

TEN

TECHNOLOGY IN THE BIOLOGICAL WORLD

TERUHIKO BEPPU

THERE IS one pattern to the progress of any branch of science: it begins with observation, then proceeds to experimentation and systematization, and gradually it is applied exhaustively to technology. However, biological science has been unique in this respect. Long before anything scientific was known about microorganisms, human beings had employed them to enjoy the fruit of their labours, for example alcohol. In this sense, biotechnology is not at all a recent invention, but actually one of the oldest of human technologies. The physico-chemical approach of modern biology has revolutionized this old science and it is even thought today that human genes will be manipulated artificially in the future. There is therefore general interest in the essential features of biotechnology; exactly what it is and how far it can go.

Japan has a long tradition of sophisticated fermentation technology. Thus it is no wonder that it is uniquely strong in modern genetic engineering where microorganisms are utilized. Professor Beppu of the University of Tokyo is one of the representatives of this aspect of biotechnology in Japan. He has searched the vast world of microorganisms for those unique functions which work to our advantage.

In this interview, he discusses what makes biotechnology unique among other technologies, and what kind of insight it can give us.

———

'Could I ask about your background?'

I was born in Tokyo. My father is an artist, I suppose one of the oldest doing Western-style oil painting in Japan. He moved to Paris when he was young, during the period of the *Ecole de Paris*, but then he had a preference for Italy and moved to Venice, where he decided to stay. That was at the beginning of the Showa[1] in the 1930s. He continued to live on his own in Venice and left me for the most part at a loss. I recently managed to persuade him to return to Japan, now that he is eighty-nine years old.

---

[1] The period between 1926 and 1989.

'So do you feel a kinship with artists?'

I think I know a bit about what artists do; it seems to me that the process of painting a picture and that of research have something in common. I have often said that one of the main keys to research is how to make an image of an object, and the emphasis I place on this may be influenced by my father.

'Then you didn't have much opportunity to see your father?'

That isn't altogether true. He sometimes came back, though he seemed to be uncomfortable in Japan, and would return to Venice again. I also often went there, so I could easily guide you around the narrow streets of Venice.

'I suppose your mother must have been lonely?'

She died soon after I finished my postgraduate course.

'Were you interested in biology when you were a child, or did you aim at another field and happen to get involved in it?'

I liked biology, but then I remember myself as a child who liked science in general. There are people who liked biology from the beginning and studied insects, for example particular butterflies, but in my case, I did not have any special interest in the biological world.

'Did your experience of the war during childhood affect you?'

I do remember things like being shot at by machine guns from a carrier-based plane during group evacuation, or meeting a few people who had escaped from the bomb on Hiroshima to a remote place really deep in the mountains in the district where I lived at that time. Among people of my own age, for example the novelist Kenzaburo Ohoe, it is said that the fact of having so many lines of textbooks blacked out was a great shock when they were in junior-high school. However, I seemed to have been rather an obedient child and followed what my teacher said without thinking deeply into the meaning of this.

'This was because, in the post-war period, there were certain things we weren't supposed to know.'

Yes, but I suppose I managed to get over it.

'What was your motive in deciding to go to Tokyo University?'

In my day, it was like heaven compared with the current ordeal of examinations. I liked world history and read a lot of books about European history. I felt I was studying for examinations by turning historical events into detailed chronological tables, but of course it wasn't all study, though I somehow managed to scrape a pass.

'This was the time when the TCA cycle (Tricarboxylic Acid Cycle) had just been discovered.'

Yes, the basics had been established by Krebs. But, for example, what sort of prosthetic molecules are related to enzymes that carry out oxidative decarboxylation was being researched in much the same way then as recent oncogene research.

'Eventually, Watson and Crick's DNA model appeared, didn't it? You seem to have studied during one of the revolutionary periods in the history of biological science.'

I graduated in 1956, just when Watson and Crick's model was presented, but I knew nothing about it. In Japan, for a while, this model was not even taught in postgraduate lectures. We had no conception of what would happen in the future.

'When you started your postgraduate course, what subject did you choose?'

When I joined my laboratory, one of the senior researchers had just completed his thesis on 'Taxonomic research into Penicillium originated in Japan' on which he had worked for about ten years. In the lab, there were about ten thousand strains of penicillin fungi which proved the basis of his taxonomical research. I was given the task of analysing what sort of organic acids penicillin can be produced using paper chromatography, which was the most advanced technique at the time. But you see, there were these ten thousand strains, although it was suggested that about two thousand would be taken as typical of the collection. The task left me feeling quite desperate.

'It sounds like a wild-goose chase.'

So to overcome my desperation, I began by selecting the two strains that had the longest and the shortest taxonomic names on the list of two thousand. I then tried them out with paper chromatography, and out of pure luck discovered that the one with the longest name produced a totally unknown organic acid! When I identified this organic acid, I discovered that it was a stereo-isomer of an isocitric acid which belongs to the TCA cycle. Isocitric acid has two chiral centres and four optical isomers, and I discovered this unknown one was an L-alloisocitric acid. It had never been produced naturally, and what is more, it had a remarkably high yield of 90 per cent against the sugar consumed. That was my first job.

'Didn't you feel like joining a company?'

No I didn't. At the time I graduated, Japanese industry was still only at the start of recovery after the war and it was very difficult to pass entrance examinations to get good jobs in companies. As I said I got into university without doing much work and I didn't want to take another entrance examination.

'I think that it was also the time when the concept of biotechnology had just begun to appear.'

It would be truer to say its first wave had begun to appear, and that this was concerned with fermentation and not genetic engineering. The discovery of penicillin was one of the great breakthroughs in this sense of the term. As you know, one of the earliest technological achievements was the making of alcoholic drinks, which was a form of biotechnology using microorganisms. In this sense, microorganisms have long been the friends of mankind; biotechnology did not just fall from the sky along with the discovery of penicillin and the double helix.

'What was the significance of the discovery of penicillin?'

After the discovery of penicillin by Fleming, various antibiotics were discovered and developed one after another. Before penicillin, there was sulfonamide, which is a structural analogue of p-amino benzoic acid, that is a component of a vitamin, folic acid. The concept of anti-metabolites arose from this, and the concept of so-called 'drug design' had already appeared at this time. Penicillin, on the other hand, is a product that happened to be discovered apart from the fashionable anti-metabolite theory, and the majority of other antibiotics which were discovered after that were all the result of random screening.

'So the development of penicillin was less "sophisticated" in its methodology?'

Yes. The development of sulfonamides seems to be logical whereas this is not the case for antibiotics. After the mechanism of penicillin action was explained, however, people recognized that the shape of its molecule closely resembled the structure of two D-alanines linked together. Thus came about the unexpected discovery that penicillin blocked the synthesis of bacterial cell walls.

'You mean it must have been impossible for us to "design" such a structure as complicated as penicillin from the beginning?'

Exactly. It was only through random screening that we were able to discover it in nature. I find this quite interesting. Anyway, the discovery of penicillin created a new trend in which novel microbial strains were introduced into traditional fermentation technology. I regard this as the basic cornerstone in the development of modern biotechnology.

'The significance of research into penicillin is that it is an example of the need to go beyond basic research, testing out alternative applications. I think this is a very good example of a discovery in applied research leading to a new theory in pure science. The history of science is full of such cases. A famous example occurred in the Sony laboratories, where the tunnel effect was discovered by chance when they were trying to develop new transistors. Another example in biotechnology is amino acid fermentation, where Japanese research is especially strong. This was discovered about the same time as the rise of antibiotics, and provided a basis for recent biotechnology in Japan. It seems that this has come about as a result of rational planning. Could you tell us more about it?'

The history of the development of amino acid fermentation is also very interesting. Originally, L-glutamic acid, one of twenty amino acids which make up natural proteins, had long been used as a flavouring additive in Japan. It was produced by hydrolysis of gluten, a wheat protein. Now I will tell you a curious story about research into L-glutamic acid fermentation. Glutamic acid is an important intermediate for protein synthesis essential for all creatures. A typical example of fermentation up to then was ethanol fermentation by yeasts.

'This was one of the main concerns of classical biochemistry, wasn't it?'

Yes. This is a process where a reduced compound, ethanol, accumulates as a result of electron transfer from the coupling oxidation reaction of the substrate, glucose. It was unthinkable, at that time, that an intermediate essential for protein synthesis or building-up bacterial cells could ever be accumulated by microorganisms. This was in around 1955.

'It seems quite irrational then to imagine that such an event could really happen in the biological world?'

Yes, in principle. At that time, when I started work in my lab in Tokyo University as an undergraduate trainee, one of my classmates was asked by Professor Kinichiro Sakaguchi to look for a microorganism which would accumulate L-glutamic acid. We thought that from the viewpoint of the theory of modern biochemistry, the intermediates of protein synthesis couldn't possibly be accumulated, that he was a poor thing to be left with such a project! We teased him about it over a drink. And then it was actually realized by researchers in Kyowa Fermentation.

'A microorganism that efficiently accumulates just the intermediate of protein synthesis must have been a paradigm change.'

Yes. At that time, young students like us listened to lectures about the most advanced biochemistry of the time. These were the things that now appear in high-school science textbooks, such as the TCA cycle, which at the time seemed quite a fancy topic. From what we had learned, we argued that glutamic acid could not be accumulated, and yet a strain which did this was actually discovered.

'Against all rational expectations?'

Then, once it was discovered, we found the mechanism of glutamic acid accumulation quite interesting. It was due to a very characteristic alteration of the permeability of the cell membrane, leading to leakage of the glutamic acid produced inside the cells. Maybe it also proves the thesis that, if we look at the biological world only from the viewpoint of the basic theory of the period, it often happens that we cannot achieve a breakthrough. In this sense, in our field, it is often the case that a theoretical breakthrough comes after the discovery of new phenomena by random screening.

'How did the research into amino acid fermentation develop then?'

After that strains were developed which produced different nutritionally essential amino acids, such as L-lysine, from the original glutamic acid-producing strain. The knowledge gained from basic research into the control of amino acid biosynthesis contributed a good deal to this development. At this point our research took a very theoretical path. However, I must emphasize that it was the quite empirical research we did at the very beginning that brought about the real breakthrough. I suppose this was instructive for us.

'People tend to think nowadays that biotechnology is so advanced that we are able to "design" biological functions quite easily. But you are saying that at the real cutting edge there's always some element of blind research?'

Design is of course important, but it is wrong to emphasize it too much. We know of probably less than one per cent of the varieties of microorganisms existing now on the earth.

'It seems that large-scale screening has been going on in the American bio-industry.'

To my knowledge this is true, but in some ways it is not. In the USA they put much emphasis on rational design, and even when they do random screening, they try to do it automatically by using robots. I think this is too mechanical an approach. I'm afraid they sometimes don't have the right attitude to make use of assay systems based on personal experience, expertise, and observation.

'So their blind research is somewhat different from yours.'

Let me take an example. We discovered a class of antitumour agents after examining with a microscope the effects of diverse agents on the morphological changes of tumour cells.

'That would be impossible with robots! Are the methods you use unique to Japan?'

Maybe, to a certain extent. But in my case, I have inherited the traditions of the lab and have consciously made efforts to make this characteristic of our way of working whatever goes on in other labs.

'What are these antitumour agents? Are they unique?'

Yes, they are. They kill tumour cells selectively and preserve normal cells. Such high selectivity is obviously most desirable in anti-cancer drugs. I believe this will open up new possibilities for the future chemotherapy of cancers.

'Could you describe other works being done in your lab?'

One of the most important areas of research is the protein engineering of the milk-clotting enzyme, chymosin, which is necessary for making cheese. Cymosin is a proteolytic enzyme which is obtained from a new-born calf's stomach. This enzyme, in spite of its lack of proteolytic activity, has a strong milk-coagulating capacity. Because of this, it has long been used as the first step in cheese production. After World War II, however, after the price of cattle went up, the price of beef soared, especially in the States. People now preferred beef, so it became uneconomical to kill calves. They were allowed to grow up for meat production. As a consequence, the number of calves slaughtered decreased sharply and so the amount of chymosin produced was reduced.

'So something had to be done.'

When chymosin production fell world-wide about thirty years ago (it was at the time of our former professor), we tried to find an alternative enzyme for chymosin from microorganisms. After a series of random screenings, we discovered that a kind of filamentous mould called *Mucol pusillus* produced an enzyme that could be substituted for chymosin. This enzyme now accounts for half of the world's cheese production.

'Along with sodium glutamate, the discovery of chymosin is something that Japan can be proud of. How has the use of this substitute enzyme been developed?'

At the time of its discovery, genetic engineering was just beginning to emerge. It is important to take advantage of microorganisms in order to produce alternative enzymes when real ones are in short supply. On the other hand, it would be much nicer if we could generate by genetic engineering the microorganism which produces the real enzyme. Then we might go back to those happy days when we could use chymosin without affecting meat production. As a result, we researched into the production of chymosin by using genetically-engineered microorganisms. We actually succeeded in cloning the chymosin gene from a calf stomach for the first time ever and established a system for producing large amounts of chymosin using *E. coli* cells.

'What is the present target for research into chymosin?'

Creating chymosin with genetic engineering has been a major area of research for a long time in our lab. As a result, we now have an efficient system for producing chymosin by genetic engineering. We can now manipulate within this system and change a specific genetic code in the chymosin gene, a widely used method called 'site-directed mutagenesis'. It's a powerful technique which converts an amino acid from one position in a protein to another.

'So this is a sort of second phase in chymosin engineering?'

If it goes smoothly, we should be able to improve artificially the property of chymosin as a milk coagulant. It will be of great benefit. Also, through such research, we are aiming to elucidate the catalytic mechanism of enzymes; in this case, not only of chymosin itself, but also a group of enzymes related to chymosin.

'Because what you are getting now is not the original chymosin itself, but varieties of it, types of artificial enzymes created by you?'

Yes, this is the natural course for the research to take.

'And what will be its outcome?'

Recently, by site-directed mutagenesis, we generated a mutant enzyme which possesses a milk clotting activity which is three times higher than with a natural enzyme.

'So this is one of the real gains from genetic engineering.'

But there are more scientific implications. As you may know, proteolytic enzymes are classified into four groups. Chymosin belongs to a group called aspartic proteinase, of which the well-known digestive enzyme, pepsin, is also a member. Renin, which belongs to the same aspartic proteinase group, is an important enzyme which relates to blood pressure control; and the HIV proteinase that AIDS itself makes, which is indispensable for the viral growth, also belongs to this group. Interestingly, it became clear that all the members of this group possess a very similar three-dimensional structure. Thus, what has been discovered in chymosin could well be true for other important enzymes as well.

'So your research on chymosin has taken on strategic and general implications.'

Yes. If we managed to find a way to design an inhibitor specific to these kinds of proteolytic enzymes, we might be able to produce agents for hypertension or AIDS therapy. Inhibitors for renin have already been used in this way, and protein engineering will contribute a good deal here. Thus, by studying chymosin protein engineering, new areas of research might be opened up, for instance, the structural design of specific inhibitors of renin. This is one of the things we are doing in our laboratory.

'This is an example of the development of strategic research employing what might be called biotechnology in a narrow sense, or genetic engineering, based on materials discovered by random screening.'

Yes, it is quite remarkable that protein engineering of this sort is now possible.

'The interesting thing about the protein or peptide world is that, for example, different activities emerge merely by replacing one amino acid. A famous example is sickle cell anaemia which is often seen in north Africa near the Mediterranean. In this case, a completely different expression occurs just when glutamic acid at the sixth N terminus of the β chain of normal haemoglobin is replaced with valine. I think this sort of possibility has marvellous implications.'

Along with the advent of genetic engineering, a lot of biologists rushed into DNA research for a while, but now the importance of protein is being recognized again.

'Is it still impossible to design proteins freely?'

I would draw your attention to the fact that even a small protein consisting of 100 amino acid residues possesses an enormous number of varieties after amino acid exchange. There could be as many as 20^{100}! To overcome this devil's number, we need to be able to predict which residues should be exchanged. But the trouble is that our knowledge of structure-function relationships in protein is still very meagre.

'So though you know something interesting could be done, you do not know what that something is. Is this because some theory is missing?'

Yes, I think so. For example, we are virtually missing an electron theory of the catalytic function of proteins. We will probably have to wait for more random results before we can establish this theory.

'So there is a relationship between technology and science, as we can see from the discoveries of penicillin and superconductivity?'

These are good examples of technology leading science. It is a myth that we always move from pure to applied science. A true investigation begins only when we confront an unexpected situation. It is only half true that science always precedes technology.

'Even in Europe, where this attitude prevails, chemistry, for example, began solely from a form of technology called alchemy.'

This is especially true in biology.

'What is the main reason for this?'

There are so many things we still don't know. It might also be true of physical sciences such as pure physics; in the case of biology, what is still unknown is really enormous. I have talked about the importance of random screening but this wouldn't be true if the world wasn't full of unknown microorganisms. Even if, say, half of the microorganisms that exist had been discovered, it still wouldn't be possible to find one easily at random. In fact we have probably only discovered a very tiny proportion. In this sense, the main characteristic of biology lies in its range of possibilities, in other words, there is so much that is still unknown.

'So you are saying that biology still needs to define its own objectives!'

Well, that may be an exaggeration. Let me put it this way. There have been a lot of theoretical definitions of how to understand the essence and characteristics of living organisms; for example, as intellectual machines with a self reproducing ability; as systems pumping out entropy; or as systems with feedback control through metabolic processes. Each of these is important, but I think another fundamental characteristic is the diversity of species.

'Surely this has been acknowledged by natural historians.'

But it is sometimes neglected by molecular biologists. It may not appeal very much to them, because it is a very old view originated by Aristotle. But it is very important to emphasize it today. The huge industry of amino acid fermentation could not have been achieved without random screening which basically works from the principle of diversity. In the field of applied microbiology, diversity of species is the essential prerequisite for technological development. Furthermore, a new phenomenon could be discovered based on the diversity of species, out of which a general biological principle might be revealed. So in both pure and applied science, we cannot neglect the diversity of species.

'In the book *What is life?* written by Schrödinger in the 1940's, it is stated that, from the thermodynamics point of view, the system which includes living organisms is moving automatically towards a state of increased entropy. So organisms must have some mechanism which pumps out entropy. In order to do this, they have to establish order. For example, protein has a highly ordered structure. I remember this being a great shock to me at the time.'

It is true that this kind of theory was very surprising to biologists whose main preoccupation was the diversity of species. Since then, this theoretical or physical approach to living organisms, or to life itself, has dominated biological thinking. Out of molecular biology the central dogma arose and the unity of life was established: genetic codes are universal from *E. coli* through to elephants. This is important in itself.

'You think the pendulum has gone too far in the other direction?'

Well, talking about biotechnology, I would interpret the trend in this way. Genetic engineering is based on the unity of life and it has created a great wave of recent biotechnology. On the other hand, technology based on the diversity of life when amino acid fermentation was discovered is still growing as a substantial part of biotechnology, and I consider this as another important aspect. The future of biotechnology depends on whether these two streams can be successfully merged.

'Out of the whole world of microorganisms with their great diversity and complexity, human beings just happened to discover penicillin. But do we really know what antibiotics mean to the life of microorganisms?'

This is often discussed both as a joke and as a serious subject. Certainly they don't exist in order to cure human diseases! But we have not yet found any sufficient proof that microorganisms produce them for their own benefit.

'I understand you have made a discovery which is related to this topic.'

An interesting feature of antibiotic production by microorganisms is that it is controlled by some kind of hormone. The well-known streptomycin is produced by a species from a characteristic group of bacteria called *Streptomyces*. We found that it produces a low molecular weight hormone called the A-factor which switches on the streptomycin production by itself. More strikingly, the same hormone also induces morphological differentiation to form spores.

'What does this suggest?'

It's therefore certain that antibiotic production controlled precisely by hormones is related to the cellular differentiation of bacteria, though we still can't explain what benefit this has.

'So is it possible that they are simply just producing them?'

One idea would be that low molecular weight organic compounds evolved autonomously much earlier than the appearance of proteins on the earth. In a way, some organic molecules may be seen as fossils even before life evolved.

'But today, all organic molecules are made from protein, aren't they?'

Now proteins work as catalysts in the cells, but it is possible that organic compounds played the role of catalysts long ago. Anyway, this is one of the most interesting ideas.

'"RNA World" is another such idea, isn't it?'

Yes, the so-called central dogma is induction from the current life-forms. So it's quite possible that the origin of life may have been more chaotic.

'So even among the diversity of microorganisms, or life as a whole, is there a chaotic aspect which is difficult to explain?'

Not really. The standard theory is rather that each characteristic of a living organism corresponds to a niche, or environmental constraint, which explains its *raison-d'etre*.

'You mean the variety of living organisms corresponds to the possible combinations within a micro-environment?'

For example, we discovered that microorganisms which live in absurd conditions, in extremely alkaline or hot environments, possess unique alkalophile or thermophile properties.

'So whatever environment there may be, it can always support life to a certain extent.'

Yes, to some extent. I will describe another interesting aspect of microbial ecology here. A long time ago, we conducted a series of screenings to find thermophilic microbes which might produce useful enzymes. Soon we found several samples which indeed contained such organisms, but we couldn't isolate these strains at all. We gave up for several years. But eventually, it turned out that the bacterium we were looking for had a unique character: it couldn't grow by itself but needed another specific bacterial strain to coexist with it. This phenomenon is called symbiosis. This discovery, as a matter of fact, has posed a number of interesting questions about current methodology of microbiology. But here I would remind you that symbiosis is one of the major strategies by which microorganisms establish their niches in the ecosystem.

'Do you think there are any environments on the earth where no living organisms can dwell, if you take these microbes into account?'

No, so although living organisms may be different in terms of complexity, they cannot be considered superior or inferior. Today, the destruction of the 'environment' here refers only to humans or mammals. No matter how bad the environment becomes, so bad that human beings or rats or cockroaches may become extinct, there will be a microorganism that will live on or even flourish in it.

'It sounds quite a cheerful possibility for the future! Now, can we return to us gloomy humans; could you tell me what you think is special about Japanese biotechnology?'

As I said earlier, the attitude and approach of 'hand-made' screening is what the Japanese are good at, whereas the English, for example, are good at theories of structure using X-rays, where the Japanese are trailing far behind.

'Maybe the Japanese have a rather different attitude to research?'

Your question reminds me of something interesting. In a temple in Kyoto, there is a monument for microorganisms called the 'microbial gravestone'.

'Why was it made?'

I don't really know. Maybe this is a memorial for the microbes which have served for the benefit of human beings. Many Japanese universities have memorial tombstones for their laboratory animals.

'In the West, people would protest about using animals for experiments.'

The attitude of the Japanese, I think, would not be seen as 'sentimental', but probably is rooted in a sort of botanical animism that they do not want to stay too far away from living things.

INDEX